Real World IT Projects

A NON-TECHNICAL GUIDE
TO HELP YOU SUCCEED

Steve L. Pinckney

Conceptia Publishing
HOUSTON, TEXAS

Book Layout © 2018 BookDesignTemplates.com and Christopher Derrick

Editor: Lona K. Neves

ISBN 978-1-7326297-0-7
Published by Conceptia Publishing - CS
12941 North Freeway, Suite 633
Houston, Texas 77060

www.conceptia.com
Editor@conceptia.com

To my mother and father,
who taught me to bloom where planted.

And to Jeff and Justina,
who taught me how to help others do the same...

Contents

Foreword

I have worked in Information Technology (IT) for over twenty-eight years. I began in the industry in the late 1980s as a software coder and database administrator before arriving at the door of project management in 2002.

During this time, I have watched technology shape and reshape the business environment. From architectural changes such as centralization to decentralization (and back again) to the advents of mobile and cloud computing, technology has impacted nearly every aspect of business in almost every industry. Now, there are even more advances in technology changing the technology landscape soon: cybersecurity, big data, machine learning and robotics process automation to name the most popular. And just as before, with these changes will come additional advantages...and challenges.

One of the biggest challenges many organizations are now facing is how to get IT and the business to work together so that the maximum value is extracted from these technologies.

IT can no longer simply "check the box" on user engagement. Nor can the business continue to remain aloof about how IT teams work together to build the tools needed by the business. Instead, each must stretch into the other's bailiwick, understanding the abilities and limitations of each. The traditional view of IT as "Service Provider" and the business as "Customer" must give way to one where the organization is seen as the customer.

So, with so many books about project management on the market, why read this one? Quite frankly, because it is not a book on project management. *Real World IT Projects* is a primer written to help someone with little or no technical knowledge quickly get acclimated to working on IT projects and with the roles the projects are comprised of. It answers all the relevant questions that someone who might be starting out or who is unfamiliar with "project basics" might ask.

Steve has delivered a wealth of information in this book on IT projects in a clear and open style that I enjoyed. It contains need-to-know "practitioner" definitions without becoming dry or overwhelming. It also introduces the roles of the project team in a friendly and straight-forward way. I recommend reading these chapters thoughtfully. In the past when dealing with interpersonal problems on a project, I often found the source of the issue to be rooted in misunderstood roles and responsibilities.

Because he has built this book on experience that can be applied to any work environment, anyone who is part of a project team will find it useful regardless of the industry or sector. With its easy to follow conversational style and use of everyday examples, Steve has provided a concise and practical guide that can serve as an invaluable guide to new entrants to this space.

A good primer should exhibit the following three characteristics:

1. Simplicity of concepts
2. Clear understandable language
3. Most importantly, a knowledgeable writer who has extensive experience with the subject

I believe that Steve has delivered on all three. Also, the presentation style and formatting of the book make it useful as a great desk reference that you'll want to refer back to over time.

Learning sets you up for success in the long-term and reading Steve's *Real World IT Projects* will add to your arsenal of knowledge. When I look back on own my journey as an IT professional, some of the missteps that I made could have been easily avoided had I been provided early in my career a basic primer like the one Steve has written.

Carla Fair-Wright, PMP
July 1, 2018

CHAPTER 1

Why This Book

Tis is not a book on project management. I do not share "Umpteen ways to deliver project success!" with you. Nor do I discuss terms like *earned value, critical path*, or *resource leveling* (although I do touch a bit on *resource allocation* in Chapter 4.)

No, this is a book about the fundamental components of an IT project. Specifically, what people mean by the term *IT project*. It is a book written for anyone who suddenly lands in the organized chaos known as an IT project, but who has no idea what an IT project really is. You're intelligent and you're adaptable, but you may not know enough about the landscape to know *how* to adapt. You have many questions. What's expected of you? How are you supposed to know who to go to for what? And what are all these people doing here? Perhaps one of the following scenarios sounds familiar:

❖ You graduated from college a year ago and work for a large oil and gas company as an accountant. One day, your boss tells you that you will help roll out a new software module as a Change Analyst. You will report to Sharon, who has been appointed the Change Manager on the project. Neither of you have any experience as a Change Manager on an IT project but are expected to begin immediately.

❖ You are an administrative assistant on a project that is currently entering its final phase before completion. As team members begin to roll off, you are asked to backfill them. You've been supporting the project for nearly two years,

1

but you don't really know what all these people are doing day in and day out. You raise this as a concern with your boss. "You're smart," she says. "You'll figure it out."

❖ You are the office administrator for a small software development shop. You believe that you are extremely under-utilized and capable of more; however, you don't really know what the company is doing on a day-to-day basis, so you are not sure how and where you can add more value.

❖ You have been a Software Developer in the finance sector for over ten years. You realize one day that, for many of the projects you're on, the reason they are having difficulty is because the Project Managers do not have nearly the amount of experience you have. However, you don't feel you understand enough about project management to even begin running a project.

Or maybe one of these applies:

❖ You are manager who was recently assigned to manage a team of IT staff and want to better understand their day to day responsibilities.

❖ You are a coach who believes that you can better serve your clients if you understand the world that they inhabit.

This guide was designed to provide you with a good foundation of knowledge to help you quickly understand the world of IT projects. In just a few hours, after reading this book, you will have a more holistic understanding of IT projects and develop an appreciation for the complex system that an IT project team is.

CHAPTER **2**

Getting Oriented

(How This Book Is Structured)

T
he first section of this book gives you a foundation for what a project is.

I don't cover much theory or advanced techniques in project management (that's for another book). Instead, I focus on telling you only what you need to know right now:

❖ What a project is

❖ Common phases that projects have

❖ Types of projects you may be assigned

❖ Typical project deliverables

❖ Foundational information about project structures and management

I dedicate the second section of the book to the roles that make up an IT project team[1], so that you learn about each role's

❖ Purpose

❖ Approach to doing work

❖ Key deliverables

❖ Common challenges and concerns

---

1. Throughout this book, I capitalize these IT project role names as proper names to draw your attention to them and emphasize their importance.

The purpose of this section is not to train you to do the jobs of the roles described, but instead to help you understand the expectations of your job on the team and give you valuable insight into the responsibilities, challenges, and main concerns of every other member on your team[2]. You learn about the key processes and major deliverables that each role either leads or supports and how they all interact, contributing to an effective IT project team. This section gives you a better understanding of the different project roles and how you can engage with them to achieve the best results, both in the short-term and the long haul. You gain knowledge about the pitfalls to avoid and the inroads to make with various team members, based on what's most important to them and their roles in the team.

Along the way, I also cover key topics and concepts that help round out your understanding of IT project dynamics. These topics include components of an IT system, key terms and concepts, and more.

Finally, the book ends with a mock project, in which we explore what happens during each phase of a simple, fictional project and how the various roles interact with each other.

Now, fix yourself a cup of tea or coffee, settle back, and imagine that it's Monday morning. We are sitting in a conference room with a whiteboard, and I'm giving you your first overview of what's going on here.

A Note About Gender Before We Get Started

I believe any gender can perform any of the roles in this book. Throughout the text, I use the pronouns "he" and "she" merely as a device to simplify the language.

---

2. Visit http://www.conceptia.com for details about upcoming toolkits and training on how to perform some of these IT project roles.

CHAPTER 3

Let's Start With What a Project Is (and Isn't)

So, What's a Project?

Since you are working on a project team, a good place to begin is identifying what a project is and, just as important, what it isn't. The Project Management Institute (PMI)[3], defines a project as follows:

> *"A temporary endeavor designed to bring about a new product or service."*

That's great, but what does it mean?

Let's break it down and illustrate with a few examples.

- ❖ **"A temporary endeavor"** means that the activity is not forever and ongoing. It has a definite start date and stop date.

- ❖ **"Designed to bring about a new product or service"** means that the purpose of a project is to create something that did not exist before. One example is creating a product that the company will sell (like a new car model or software product) or a new service that the company will offer (like an oil change service or payroll service).

---

3. PMI sponsors the Project Management Professional (PMP®) certification and is the de facto standard for project management in the US. Although there is no one "right way" to manage a project, PMI attempts to identify what areas a Project Manager should be knowledgeable in, as well as the phases projects go through.

Read the following table for a few examples of items that represent and do not represent projects. As you read the "Not a project" items, ask yourself, "When would this end?" and "What is being created?"

Not a project	Is a project
Marketing a company's products	Implementing a new marketing software platform for the Marketing Department from January 1 through March 30
Selling cars	Designing a new model of car to sell
Going to college	Selecting a college to attend or obtaining a bachelor's degree in computer science

So, in plain English, a project is something that you temporarily undertake to create something new. It is not something that you do on a regular basis, indefinitely—that would be an operation.

It's as simple as that.

Phases of a Project

Regardless of its duration (whether one month or one year), a project goes through five distinct phases, explained in the following list. To illustrate these phases with a real-world scenario, I use an example of moving from one house to another.

1. **Initiating:** The administrative phase. Creating the project charter (something like a birth certificate for projects), identifying the objectives for the project, and securing funding.

   ▪ Initiating the move: You begin by signing agreements with a Realtor to show you available houses and represent you during negotiations. Once you find a house, you fill out forms to apply for a loan and have the home inspected.

2. Planning: Thinking holistically about what needs to happen from beginning to end, breaking it down into chunks or major blocks of work, and then breaking down these chunks even further to the task level. Tasks are items that you can estimate in terms of hours—generally between two to eight hours. For example, if you are building a house, one major activity is "paint house," but it is hard to estimate how much time that will take. However, two tasks under this major activity might be "identify colors" and "paint master bedroom," both of which are much easier to estimate. Planning also includes proactively thinking about what could go wrong and putting a plan in place ahead of time (aka *Risk Management Plan*).

- Planning the move: While you're filling out paperwork in the Initiating phase, you start to think about what you're going to keep and what you're going to throw away as part of the move. You also start comparing prices for several moving companies. You make a list of the various utilities you'll to need to transfer or stop as part of the move.

3. Executing: Where the rubber meets the road. Pulling the trigger and commencing work. During this phase, the Project Manager "simply" ensures that there is nothing standing in the way of starting and completing tasks. In IT projects, the Executing phase often has specific work streams for *Requirements* or *Analysis, Design, Development, Testing,* and *Go Live*.

- Executing the move: Once you've purchased the home, you can begin the actual move. You call the moving company you selected and schedule the specific date for it to come and move you. The moving company arrives on the appointed date, and packs and moves your belongings. You start calling utility companies to change the address on your accounts or close the accounts altogether.

4. Monitoring and Controlling: The Project Manager tracks progress, monitors for risks (and triggers the Risk Management Plan if needed), and updates the schedule with status and new tasks that have been identified; the Project Manager constantly assesses how these things impact the completion date. This phase happens throughout the entire project.

- Monitoring and Controlling the move: You watch the movers as they pack your belongings (monitoring), ensuring that everything that is supposed to be packed is packed securely and safely. You also ensure that anything you intend to leave behind is not packed. When unexpected issues arise, you solicit ideas and suggestions, and make the best possible decisions based on the information you have (controlling). For example, if your realtor calls with a paperwork issue while the moving company is packing your belongings, you still work to resolve the issue as efficiently and cost effectively as possible.

- Another example: The packing is going too slowly, so you decide to put more resources on it to speed it up; you ask your family for help and you yourself help out. With more hands contributing to the same task, the packing goes faster. (But ask yourself, "What were those resources previously working on that might now be delayed as a result of this?")

5. Closing: The wind-down phase. In IT projects, you do things like conduct *lessons learned* sessions (what went right, what went wrong) and release resources. Several months after go live, you may also evaluate *success metrics* to determine if the project delivered the intended benefits.

- Closing the move: Once all your belongings are in your new house and your utilities are turned on, you begin to wind down the project. You pay the moving company. You file any key paperwork for use in the future—home warranty information, all your new account numbers, etc. Assuming one of your motivations for moving was being able to save money by living closer to your office, you might analyze whether you are actually saving money on gas and vehicle maintenance.

While these phases are generally sequential in nature, they often overlap. For example, some team members may start planning aspects of the project while others complete initiating activities. This is efficient since phases can get done sooner and therefore reduce the overall time it takes to complete the project.

Types of Projects

Given the broad definition of a project, there are countless types of projects. However, there are a few that you are likely to encounter at several points in your career. The following table shows the most common IT project types, along with a brief description of each.

Project Type	Description
Application Development	The project team is responsible for creating a software product from scratch. The project team must work with business stakeholders to understand their needs and the type of system they want. The project team writes code to build an application that provides what the business stakeholders need. Because the organization creates the application from scratch, this type of application is sometimes referred to as *homegrown*.

Project Type	Description
System Implementation	Just as with the application development project, the project team works with business stakeholders to determine their wants and needs. Instead of building an application from scratch, the project team looks for an existing system (sometimes referred to as a *Commercial Off the Shelf*, or *COTS*, system) that meets these needs. Once they choose a system, the team works with various organizational areas to get the system installed and tied in with the organization's other systems; this includes configuring or customizing the software to meet the business's needs. The project team works with the IT Department to develop interfaces (automated jobs that move data to and from the system) between the application and the organization's existing systems.
Upgrade	Once an organization implements a system (whether COTS or homegrown), users often request new functionality and enhancements to existing functionality. Eventually an organization needs to replace the existing system with one that contains the enhancements, and this is known as an upgrade. For an example of an upgrade on a small scale, think about the operating system that your phone uses. If you want to be able to use the newest features or the latest apps, you need to update your operating system. The longer you wait to update, the more problems you may encounter. The same is true for organizations and COTS software.

Typical Project Deliverables

Note: This section lays the groundwork for much of what you will read later. You don't need to memorize the list of deliverables presented here, and don't worry if portions of this section do not make sense to you yet. This content can be overwhelming if you are new to IT projects. We discuss many of these deliverables throughout the book.

As with buying a house, there are many documents that organizations create during a project and some of these documents are considered *deliverables*. According to Wikipedia, "A deliverable is a tangible or intangible good or service produced as a result of a project that is intended to be delivered to a customer (either internal or external)."[4] The specific deliverables a project produces depends on factors such as the type and complexity of the solution being implemented and the organization's IT processes and project management methodology. The following list contains some common IT project deliverables, with the corresponding project phase noted in parenthesis:

❖   **Project Charter** (Initiating): Establishes the mandate or need for the project.

❖   **Risk Register and Risk Management Plan** (Planning): Describes what could go wrong and what the project team will do in case it does. The purpose of this document is to proactively review the things that could happen and put a plan in place for addressing them *before* they happen.

❖   **Scope Management Plan** (Planning): Documents how the project will capture, review, and accept or reject proposed changes to scope. It also details how the project will communicate decisions to the project team and business.

---

4. Wikipedia contributors, "Deliverable," *Wikipedia, The Free Encyclopedia*, https://en.wikipedia.org/w/index.php?title=Deliverable&oldid=826464053 (accessed April 13, 2018).

❖ *X* **Management Plan** (Planning): In which *X* refers to any of the other common documents described as *management plans* that discuss how an aspect of the project will be managed. This includes (but is not limited to) Communications, Procurement, Quality, and Schedule to name several. Structurally, these individual plans may exist as separate documents or one large document with a section dedicated to each.

❖ **Project Plan** (Planning): Describes how the project will be run and managed. It is the aggregate of the individual management plans described earlier.

❖ **Project Schedule**[5] (Planning): A task list that includes who owns each task, the start and stop dates for each, and any dependencies.

❖ **Requirements Document** (Executing): Requirements describe what the product, service, or solution must do or what capabilities it must provide. For example, if a project is delivering a new digital address book solution, it might have requirements like "The solution must provide a State field for domestic addresses" and "The State field value must be a valid United States Postal Service (USPS) two-letter state code." Requirements have a wide range of styles and formats and are often captured in one or more official IT project documents. These documents can go by many names: Functional Specification, Requirements Document, Business Requirements Document (BRD), Functional Requirements Document (FRD), Product Requirements Document (PRD), Use Cases, User Stories, and many more variations.

---

5. People often refer to the Project Schedule as the Project Plan. As noted in the previous bullet, the Project Plan describes how the project will be run and managed, whereas the Project Schedule identifies who does what and when. This is a distinction that only a few truly care about, so when you hear it, just go with it, but be clear about which is actually being discussed.

❖ **Design Document** (Executing): While requirements documentation focuses on describing *what* a solution must do, design documentation focuses on *how* the solution will meet those requirements. Design documents are detailed and technical in nature. In the digital address book example, design documentation might include technical details for presenting users with a pick-list of USPS state codes for the State field, or the documentation might have technical details to validate any user-entered State values against an official list of USPS state codes. Just like requirements documentation, design documents can go by many names: Technical Design, Solution Design Document (SDD), Technical Specification, and Software Requirements Specification (SRS). For complex homegrown or customized applications, you might encounter design documentation for the overall solution along with more specific design documents for each component of the application.

❖ **Test Plan, Test Scripts, and Test Results** (Executing): A test plan describes the scope, approach, timing, and prerequisites for testing the solution to verify it conforms to the requirements. Test scripts go a step further and outline detailed steps for someone (or some automated process) to perform to ensure that the solution is functioning as designed and does not have any errors. The test plan, test scripts, and test results are all typically considered IT project deliverables.

❖ **Training** (Executing): Organizations often require projects that implement new or changed functionality to train the solution's users on how to use the solution and to train Help Desk and IT operational staff on how to support the solution. Training deliverables might include user guides, quick reference cards, how-to videos, training presentations, or instructor-led classes.

❖ **The Solution Itself** (Executing): The system the project team is implementing is one of the most important project deliverables!

What Is a Program?

At this point, you should have an idea of what a project is, what it isn't, how a project is structured, and what typically goes into starting a project. We now look at a special type of project: a *program*. Oxford Dictionaries defines a program as a "set of related measures or activities with a particular long-term aim."[6] In the context of IT projects, these "related measures or activities" are projects. According to Wikipedia,

> *"Program management or programme management is the process of managing several related projects, often with the intention of improving an organization's performance."*

One way to think of a program is as a large project with multiple, related subprojects. The subprojects are coordinated with each other in a way that delivers organizational benefits that might not otherwise be achieved if the projects are executed independently. Consider the merger of United Airlines with Continental Airlines. Clearly this was a massive change to both organizations and one that each felt in multiple areas:

❖ The team responsible for the merger needed to work with the leadership to determine which redundant groups to merge together and which to eliminate.

❖ IT systems—including reservation systems, baggage

---

6. "programme," *Oxford Dictionaries*, https://en.oxforddictionaries.com/definition/programme.

tracking systems, and catering systems—needed to be merged.

❖ Marketing materials needed to be updated to reflect the new branding globally.

❖ Loyalty programs needed to be merged into a single program, with customers' mileage points transferred into the new program.

Each one of these was likely considered a project in and of itself. However, managed together in a program, the results of each project contributed to a larger outcome: the emergence of an airline that is the sum of (hopefully) the best parts of each airline.

Components of a Program

Programs are generally complex, far-reaching, disruptive initiatives. Because of this, they often resemble smaller organizations within the larger organization, requiring their own administrative support and governance structures.

The following figure lays out a common program structure. Note that each project has its own work streams (Business Analysis, Software Development, etc.) and that each falls under the overarching program. The Project Management Office and Change Management Office (I discuss these in the next chapter) are part of the program and support the projects with methodology, training, and templates. The two offices also support the program by aggregating data from the projects, such as status, issues, and risks.

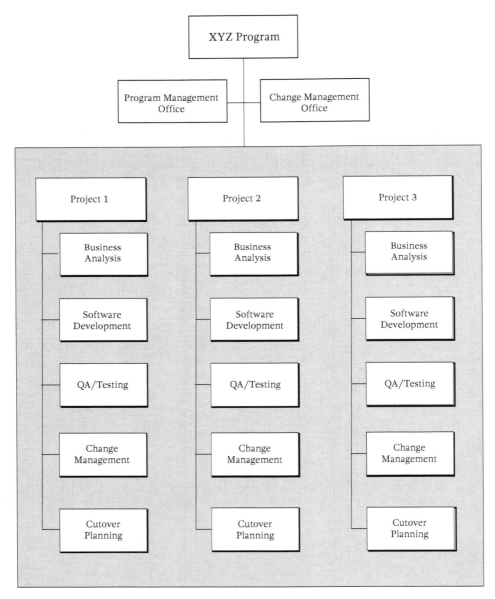

Figure 1: Components of a program

CHAPTER 4

Some Concepts You Need to Know

Before I introduce each project team member, there are a few things you may need to be aware of—nothing too critical, mind you—just a few topics that might explain comments or behaviors you may see from time to time.

Stakeholders

No book about projects would be complete without discussing the concept of a stakeholder.

Let's start with what a stakeholder is. A *stakeholder* is anyone who has an interest in the outcome of your project. This includes people who are involved in executing the project, people who make decisions affecting the project, people who support the deliverables after go live, and people who are impacted by your project.

It is important to understand that each group of stakeholders has its own perspective on the project, such as what it wants the project to deliver and concerns it has regarding the project.

Common Stakeholders

The *Project Sponsor* is the person who decides the need for the project in the first place. Just as a person can commission a piece of art for a specific event, the Project Sponsor can commission a project to accomplish one or more goals. Project Sponsors may or may not be the people paying the bills, but because they are the ones for whom the project is being executed, they often have the final say on factors such as the project budget, project scope, and how the project is governed.

The *steering committee* is a panel that convenes to provide governance over the project. By governance, I simply mean oversight. If the project needs more money to continue or to expand its scope, it is the steering committee's job to review the gathered data (usually in the form of a *decision support package* presentation) so that the committee members can decide how to proceed. The steering committee is generally made up of people who have significant experience in the focus area of the project, manage the people the project impacts, or are organizational leaders with a vested interest in the success of the project. The aggregate breadth and depth of the steering committee's experience is desirable so that the group makes quality decisions.

You often hear the term *users* on a project that implements or creates a system. Users are the people who eventually use the system to do their jobs.

Many projects have another stakeholder called the *vendor*. The vendor is the company (or person) that made or sold the software being implemented or upgraded. Different vendors have different degrees of involvement in projects, depending on the vendor's structure. Some vendors may simply provide the software and technical support and little else. Others may provide additional resources, such as personnel to assist with implementing the software (often referred to as *professional services*) and training on how to use the software.

Generally, stakeholders in an organization can be divided into two large buckets: IT and *the business. The business* is basically anyone whose primary function is not a job in information technology. *The business* encompasses the people who use the systems that IT projects implement and who execute the processes that IT projects improve. This group includes, but is not limited to users, the users' managers, and other departments that interact with the users. At an investment bank, some members of the business would be the accountants and traders. In a school system, it might be the principals and teachers who use the computer system to log student grades and attendance. In a hospital, it might be the nurses and doctors who use a computer system to enter patient data.

Project Scope

I wrote briefly about scope in the "So, What's a Project?" section in Chapter 3. Scope is actually a bit of a hot topic on projects, so I want to say a little more about it here. Scope has two components: what the project will deliver and—just as important—what it won't deliver. If a project is a map, then scope is the map's boundary or border (e.g., you wouldn't expect to see Europe in a street map of Houston). Scope is how you frame your photo when taking a picture with a camera or phone. (What are you including and not including?)

Agreeing to scope up front is of paramount importance. As a project progresses, team members and other stakeholders increase their interaction and they begin to think more deeply about the possibilities the project can bring; this makes them inclined to want to add items for the project to deliver. While on the surface this may appear to be perfectly reasonable and normal, it can wreak havoc on an IT project because projects need to have a definite end date that everyone is aiming for. If stakeholders continuously add items to the list of things for the project to produce, then it is difficult (if not impossible) to predict when the project will be completed. This issue is so common that it has its own name: *scope creep*.

However, there are times when it makes sense to increase project scope and add items to the list. Perhaps the leadership thinks it is more cost effective to do an additional piece of work in the project while the experts are in the building and readily available. Or perhaps another project is doing similar work that impacts the same group of people, so the leadership decides to move that scope of work into your project.

Maintaining control over a project and keeping its list of deliverables fixed, while providing flexibility when needed, is a fine line to walk. To do this, the Project Manager works with the project team and leadership to create what is commonly called a *Scope Management Plan*. The Scope Management Plan contains the process the project follows when someone wants to introduce an additional scope item to the project or remove a scope item from the project (usually via a *change order* or *change request* that the project governance board reviews and approves/rejects).

The bottom line is that project teams, and especially Project Managers, are wary of anything that sounds like it might be adding scope to the project. Having a solid Scope Management Plan that everyone has agreed to (or at least, pretended to agree to) lessens the tension around this topic.

Project Allocation

Project allocation is the amount of time a person is assigned to spend on a project, usually expressed as a percentage. For example, a person may be "25 percent allocated to a project." This means that in a typical forty-hour work week, that person is expected to spend no more than 25 percent of his time, or ten hours per week, working on the project.

Allocation is important for the Project Manager to know because it determines how much work a person can accomplish in a given week, which in turn can affect the schedule and go-live date.

The Project Management Office

Not all organizations have a Project Management Office (PMO), but for those that do, its purpose is to provide consistency in how projects

 ❖ Initiate, manage, and close projects

 ❖ Track and address risks and issues

 ❖ Identify and monitor dependencies

❖ Gather and report status

❖ Meet quality targets (e.g., testing processes)

A PMO is like an air traffic control tower—coordinating how projects take off, fly, and land.

A PMO provides details for things like the required process to follow and documentation to create when initiating a project. A PMO may also provide guidance on when status reports are due and the format to use when reporting status. Depending on the maturity of the PMO, it may also provide training to Project Managers to ensure they all have a certain competence level.

Finally, a PMO's primary job is providing a consolidated view of projects currently being executed to assist with decision making activities, such as budgets, resources, and costs. This consolidated view is often in the form of an integrated program plan that highlights key internal dependencies (other projects in the program) and external dependencies (other projects and events happening in the organization). Having a consistent way of managing the projects under the PMO's purview helps with this.

The Change Management Office

The Change Management Office (CMO) is still a relatively new concept, but all the cool kids are doing it. It has many of the same responsibilities as the PMO, except its support is primarily for the Organizational Change Managers and the change management efforts that are underway in the organization.

CHAPTER 5

Now, Let's Get Ready to Meet the Team

What I'm Going to Share With You

As I introduce you to each role, I share what I believe is the most helpful information about the role. My goal isn't to tell you how to *fill* that role; it is simply to provide you with enough details to make working with that person easier. My hope is that, by understanding the responsibilities and challenges of each role, you are better able to anticipate your team members' needs, provide them with information in a format that works for them, and know who to go to when you need something.

CHAPTER 6

Here's Who's (Maybe) In Charge

Project Manager

What This Person Is Responsible For

Basically everything.

This is where the buck stops. This is the "one neck to choke."

If a project is a symphony, then the Project Manager is its conductor. She helps ensure that the team plays together smoothly. She lets each section know when it's supposed to start performing and how. She sets the tempo by assigning more or fewer people to a task as necessary. And perhaps most importantly, when the project is over, she graciously bows and highlights the orchestra sections who deserve credit and appreciation.

What This Person Is Expected to Deliver

As the leader, this person is expected to deliver a lot.

The Project Manager is expected to put all the project controls in place (e.g., scope management and risk management processes), accommodate leadership's needs, lead the team, and provide project status to the stakeholders.

How This Person Approaches the Job

As part of the planning process, the Project Manager creates a project schedule, which is a comprehensive list of all the tasks that the project team must complete. Because Project Managers are seldom experts in all aspects of the project, they rely on team members to develop their own sections of the schedule. The Project Manager then works with each team (or team member) individually to identify dependencies among the various tasks. A dependency is when one task relies on another task. For example, you can't build a house until you lay the foundation. In the context of an IT project, a Software Developer cannot design and build the solution until the Business Analyst provides the requirements.

Project Managers must explicitly identify dependencies so that they know how a change to a task affects subsequent tasks. For instance, if the Business Analyst needs an additional month to complete requirements gathering, then it will be at least another month before the Software Developers can begin, which also means the Software Developers will finish later, which in turn means the overall project will finish later than planned. Without linking tasks together through dependencies, it is nearly impossible to tell how one change impacts the overall schedule.

Project Managers also consider which project activities can be done in parallel and which must be done in sequence, taking into account details such as the number of people available, the complexity of the individual tasks, how long each task takes, and any dependencies. Consider the following diagrams.

Sequential Task Execution

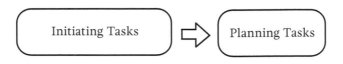

Figure 2: Tasks executed sequentially

Sequential tasks have the following characteristics, as compared to parallel tasks:

❖ Tasks begin only after previous ones complete

❖ Fewer resources are required

❖ Overall timeline is longer

Parallel Task Execution

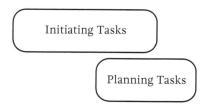

Figure 3: Tasks executed in parallel

In contrast with sequential tasks, note the following characteristics of parallel tasks:

❖ Tasks are executed concurrently

❖ More resources are required

❖ Overall timeline is shorter

Once the Project Manager establishes the project plan and the project schedule, the Project Manager must track the team's progress and report project status.

While there are many ways to report project status, common items to include in a status report are the team's progress against the project schedule, how much budget has been spent, which objectives have and have not yet been met, and any identified risks and issues for the project.

To report overall project status, the Project Manager needs to know the status of all project work streams. Have the Business Analysts completed documenting requirements? How much of the system have the Software Developers built? How is the project team doing with communications? Understanding the status of the various work streams helps the Project Manager determine overall project health and whether the project will deliver the solution on the planned go-live date. Since status is such an important part of a Project Manager's world, the Project Manager often asks the various work streams' leads to provide their own status reports by specific due dates.

Common Challenges

Getting Up to Speed Quickly

One of the first challenges a Project Manager faces is understanding what the overall project goals are. Why did the Project Sponsor commission the project? Why now? What does success look like? After understanding the big picture, the Project Manager may then need to spend some time learning about the industry or subject matter of the project. Finally, the Project Manager needs to learn about the organization—its culture, who's who, and how things are done.

Balancing the Needs of Everyone

Because Project Managers are heavily concerned with scope, they must walk a fine line of maintaining course while providing flexibility to slightly alter the course along the way. It is a delicate balancing act that can result in a Project Manager being seen as a hero to some and a villain to others.

Project Managers must also be diplomatic, especially when negotiating the various stakeholder groups' conflicting needs or wants.

Working Within the Constraints of Their Authority

While Project Managers may be in charge of their projects, this authority only carries them so far; they must still work with other areas of the organization to get what they want. For example, they may wish to utilize a well-known software solution for tracking and managing project issues, but rather than simply purchasing the software for their teams to use, they may need to work with the Procurement Department, which may have its own idea of which software should be used (based on existing licensing agreements and favorable terms provided by existing vendor relationships).

Another example is that project team members often do not have a direct, employee-supervisor reporting relationship with the Project Manager, which constrains the Project Manager's ability to unilaterally adjust team members' priorities or allocation to the project (e.g., the Project Manager might want to have Joe on the project full time, but Joe's supervisor can only spare 25% of Joe's time for the project).

Knowing the True Health of the Project

Project Managers can find themselves managing large teams specializing in subjects that the Project Managers themselves know little about. This means that they must trust their teams while finding ways to confirm that the project is in overall good health.

Program Manager

What This Person Is Responsible For

Like a Project Manager, a Program Manager must also balance various stakeholders' needs with the organization's needs. However, unlike a Project Manager, a Program Manager focuses on how well *all* the projects are working together to meet the program's objectives.

In some ways, the Program Manager is like the mayor of a small town, responsible for reporting various projects' statuses, reporting the overall program's status, and getting support for the program in the form of funding, personnel, or other resources. Also, like a mayor, the Program Manager's work is as much about politics as it is the business.

The Program Manager must work with, in, and around the various power dynamics at play to achieve the program goals.

Seen from different perspectives, Program Managers are also responsible for the following:

❖ From the organizational leadership's perspective, they are responsible for delivering organizational benefits and a clear roadmap explaining *how* and *when* each will be delivered.

❖ From the perspective of the Project Managers for the various initiatives under the program, they are expected to provide support by engaging the upper levels when needed.

❖ From the PMO's perspective, they are expected to provide a clear mandate on what the PMO should deliver and how they should support the projects.

What This Person Is Expected to Deliver

In short, the world on a silver platter.

Whereas the Project Manager is responsible for delivering *things* like services or products, the Program Manager is responsible for delivering *organizational benefits*.

When I say *organizational benefits*, I am talking about things that are good for the organization or that give the organization an advantage over competitors. For example, a clinic that acquires an MRI machine and installs the necessary infrastructure to support the machine can offer more advanced diagnostic services than previously possible. This in turn results in a benefit of *increased revenue* for the clinic. It also allows the clinic to begin servicing additional types of patients, perhaps also resulting in a benefit of *increased market share*.

As another example, consider the increased use of drones that the military, oil and gas, and other sectors use to deliver benefits to their respective organizations. Real estate agencies have begun using drones to capture aerial views of properties. In the past, this service was only available for high-end listings. Today, any agency that invests in a drone and video editing software can offer this service quickly and simply. For the agency, this represents an organizational benefit of increased revenue because, for an additional charge to clients, it can offer commercial-quality, aerial videos of their listings. Building this capability would likely require a program that includes one project to select a drone and train the relevant team members in its use, and another project to modify the agency website to display videos. The investment the agency makes in this program provides it with a competitive advantage over agencies of similar size and status. You can almost hear the sales pitch, "You can list with the agency down the street and get your property listed OR you can list with us, get your property listed, and have a nice video of your property displayed prominently on our website."

The Program Manager is also expected to deliver a comprehensive roadmap showing how and when organizational benefits will be realized. Consider a program that is rolling out an internal social media platform to an organization with offices worldwide. The platform allows users to post written articles or items they think other employees might find interesting. It also allows users to post videos if this feature is enabled, but this option requires additional servers, storage space, and other network equipment upgrades to handle the resource demands required by video posts. The Program Manager, in consultation with the project teams and Subject Matter Experts, could opt for one of several approaches:

❖ **A feature-by-feature approach**: The ability to post written articles is deployed to each region sequentially—first to the North American offices, then European offices, and finally offices in Asia / Australia. Then the ability to post videos is deployed to the same regions, in the same order.

❖ **A region-by-region approach**: Both articles and video

posts (and all the required upgrades) are deployed one region at a time, starting with North American offices, then European offices, and ending finally with offices in Asia / Australia.

❖ **A parallel approach**: The ability to post written articles is simultaneously deployed to each region in parallel, followed by deploying the ability to post videos to each region in parallel.

These are not the only possible approaches to rolling out the platform. Another approach might be department by department. That is, both features could be rolled out to all accounting departments globally, then all human resources departments globally, and so on. Still another approach might be by the levels in the organization, first to the executive leadership, then to middle management, and then to the remaining employees. The approach selected is ultimately based on several factors, including the complexity of the platform, the experience the team has with deploying the platform, and very often the budget.

How This Person Approaches the Job

A program is, in effect, a team of teams. As such, its overall success depends on the success of the individual teams, which in turn depends on the leadership of the individual teams.

With this in mind, Program Managers choose strong Project Managers to lead the various projects, knowing that this frees them to focus on the overall program and not the issues and details of each project.

Similarly, Program Managers work to establish a strong PMO, capable of providing the Project Managers and their teams with the necessary tools and templates to function effectively within the program. This too frees the Program Manager from details such as obtaining the status of each project or explaining how issues and risks should be managed.

Common Challenges

Defining Clear Priorities for the Program

Program Managers must obtain agreement from the organization's leadership on the priorities for the program. Is minimizing the program's cost the primary goal? Or is it getting the solution deployed as rapidly as possible? More complicated still, is getting a particular department's needs addressed more important than other priorities? Even after the leadership is aligned on the program's priorities, the Program Manager may need to revisit this agreement from time to time as new challenges emerge.

Showing Near-Term Success

Because a program's benefits may not be realized until years after the program is completed, the Program Manager may find it difficult to show success in the near term. For example, an organization may decide to deploy a software platform across the enterprise, with the belief that it will save money in the long run. These types of benefits can sometimes take years to manifest. During program execution, the Program Manager needs to continually remind the decision makers why this effort was originally undertaken and that the value to the organization is expected to be realized later down the road. Even so, the Program Manager needs something to indicate near-term success. For example, a program that is implementing a Help Desk solution might show how the average time taken to resolve customers' issues is gradually declining. Returning to the real estate agency example, the Program Manager might share customer feedback or testimonials for the initial videos the drone program produced. While the long-term goal of increased revenue may take a year or two to be reached, the program can show that customers are already responding positively to the videos, giving the leadership confidence that the decision to implement the drone program was a good one.

Keeping on Top of Required Communication Demands

Due to the nature of the role, the Program Manager is often at the center of everything. There are leaders "above" him that require updates and feedback. There are project teams "below" him that require guidance and input. There are Program Managers for other programs "beside" him who might also need to be kept abreast of his program's progress (and vice versa). Because of these various needs, it can be difficult for a Program Manager to keep up with the required communication demands.

Staying responsive to programs often takes years to be fully completed. Many changes that could affect the program can occur during this time. Here are a few examples.

- ❖ **Organizational needs**: the organization decides to change strategy, and begin making and selling vehicles overseas

- ❖ **Market conditions**: a decrease in demand for fast food or a shift in preference from gas vehicles to electric vehicles

- ❖ **Regulatory conditions**: the city requires that builders elevate homes to be prepared for expected increases in flood levels

Being responsive and adapting to these and other emerging needs is at once a crucial skill and a key challenge for the Program Manager.

Working With Higher Level of Stakeholders

The Program Manager faces many of the same challenges as the Project Manager, only amplified because the Program Manager usually engages higher-level stakeholders. It's one thing when a middle manager disagrees with your approach or how you handle a situation; it's quite another when the CEO disagrees with you.

Organizational Change Manager

What This Person Is Responsible For

In short, stakeholder adoption of whatever the project is delivering.

The Organizational Change Manager's job is to identify the various stakeholder groups; determine how each group will be impacted and if that impact will be high, medium, or low; and put an action plan in place to prepare each.

Consider the rollout of a new accounting system for an organization's Finance Department. Imagine the chaos that would ensue if everyone came in on a Monday to discover that over the weekend the organization switched to this new accounting system with no warning or advanced announcement to anyone. Imagine how caught off guard the Help Desk would be when call after call comes in, reporting that people cannot find the old accounting application on their computers and that something must be wrong.

Admittedly, this is a (slightly) exaggerated scenario to illustrate several points. As you think about the scenario, ask yourself the following questions:

1. What happens to productivity starting that Monday and for the next month or two?
2. When will the organization begin to see productivity gains and therefore begin to recoup its investment?
3. How do you think business users will react in the future whenever they hear that something might be changing?

The primary goal of organizational change management is to reduce the dip in productivity as much as possible. By proactively preparing stakeholders for a coming change, Organizational Change Managers help ensure that people are aware of a change, understand the reasons for the change, and perhaps most importantly, understand how the change affects them specifically.

After go live, the sooner stakeholders begin to use the system functionality that was implemented and follow the processes as designed, the sooner productivity can begin to return to pre-go-live levels, and an organization can begin to recoup its investment. Said another way, the sooner productivity returns to pre-go-live levels, the sooner the organization can begin to realize its return on investment.

What This Person Is Expected to Deliver

Most commonly, the Organizational Change Manager is expected to deliver a combination of the following deliverables, which are designed to prepare the various groups for a coming change:

❖ Communications – Could be in the form of emails, posters, explainer videos—anything that could be used to get a message across

❖ Engagements – Usually for specific topics and provide opportunities for stakeholders to provide feedback to the project team

❖ Training – Instruction on how to utilize the system being implemented or how to follow the processes being delivered by the project

How This Person Approaches the Job

The mental model for an Organizational Change Manager is something akin to a hill, and the goal is to get each group to the top of the hill by a certain date, passing several landmarks along the way. The hill represents the overall journey a person takes and the landmarks represent the different phases of change, beginning with awareness of a change and ending with acceptance or ownership of the change.

Organizational Change Managers develop targeted plans and strategies for each group to get the group up that hill. These plans are based on the degree to which a stakeholder group is affected. If the change is relatively small for a group, then perhaps a single communication is sufficient. However, for larger changes, a mix of communications, face-to-face meetings, and training may be required.

Consider the difference between "we are changing the application icon for the accounting system the Finance team uses" and "we are changing the accounting system the Finance team uses." For the first scenario, a simple email communication making users aware of the change, why it is changing, and where to go for help if needed is enough to prepare most users. The second scenario calls for the Organizational Change Manager to employ a mix of interventions:

- ❖ Sending multiple emails explaining the various ways the new system will affect the Finance team

- ❖ Communicating topics ranging from how users will log on to how invoices will be processed

- ❖ Organizing multiple face-to-face meetings and conference calls to give users a chance to ask questions about the system before it arrives

- ❖ Depending on the complexity of the system, organizing training to explain how to perform their jobs in the new system

It is the Organizational Change Manager's job to work with the various groups to understand what it will take to prepare them for a change, and then develop and execute plans that will do so.

Common Challenges

Being Properly Utilized by the Team

While not exactly new, organizational change management can still be considered a somewhat emerging profession, especially in smaller organizations. As a result, not many people understand the true goals and responsibilities of Organizational Change Managers, which leads to Organizational Change Managers being assigned tasks that don't seem to fit in anyone else's role. Because of this, Organizational Change Managers must often spend time explaining their role to their own team.

Not Being Involved Early Enough in the Project

Because many people do not understand an Organizational Change Manager's role, they often do not involve the Organizational Change Manager as regularly or as early as needed. The sooner an Organizational Change Manager is involved in a change, the sooner she can begin developing plans to mitigate the anticipated impacts of the change. This can mean the difference between proactive intervention and reactive intervention, which in turn can affect the success of the project.

Identifying All Impacted People and Processes

A project implements changes that impact people, processes, and systems. It's usually clear which system(s) the project needs to address. Unfortunately, it's often less clear which people and processes the project needs to accommodate.

As organizations develop and mature over time, processes and reporting lines change and these changes are not always formally documented. Additionally, people begin to develop their own ways of getting work done—ways that might not be obvious to anyone outside of their group. Because of these realities, it can take time for the Organizational Change Manager to *find* all the impacted stakeholders and processes.

The Organizational Change Manager, and to a large extent the Business Analyst, must find the additional groups the stakeholders work with and the processes they follow, and then work with them to determine how best to prepare them.

CHAPTER 7

Here's Who Does All the Work

Business Analyst

Like the Organizational Change Manager, *Business Analyst* has become one of those titles that mean many different things to different people. It has almost come to mean whatever the person speaking about it thinks it means. This is partially due to the flexible, multi-skilled nature of the role. If you have a problem, a Business Analyst most likely has tools that are not only useful for highlighting and illustrating your problem from multiple perspectives, but also for thoroughly describing a detailed solution.

What This Person Is Responsible For

If the Project Manager is responsible for "getting things done" and the Organizational Change Manager is responsible for getting the people "ready" as said things get done, then the Business Analyst is one of the key people responsible for identifying what things need to be done. According to a traditional, simplified view of the role, a Business Analyst is generally concerned with three things:

❖ How are things done today?

❖ How will things be done tomorrow?

❖ What things need to be done to get us from where we are today to where we want to be tomorrow? (This is also known as *gap analysis*.)

The Business Analyst works closely with the business to understand the need for the project's components well enough to translate these needs into requirements that Software Developers use to build products. For example, if the project team is building a ship for a client, the Business Analyst is responsible for the following:

❖ Working with the client to understand what the client currently uses, how it uses it, and how it intends to use the new ship

❖ Providing a detailed description of how the client pictures the ideal ship

❖ Documenting the ship's exact specifications, including length and cargo capacity

❖ Describing detailed processes for the ship's ongoing operation

The way the Business Analyst communicates the business needs to the project team is through a set of requirements. A *requirement* is a description of one aspect of the item being created or built. Returning to the ship-building example, one requirement might be that the ship "must be capable of holding 500 tons of cargo," or that it "must contain living quarters (each consisting of a bed, sink, and toilet) for one hundred crew." It is very important that requirements are as detailed as possible, to avoid misinterpretation by the various people who read and consume them.

But the Business Analyst doesn't stop there. A new feature, system, or process often necessitates other changes; it is the Business Analyst's responsibility to recognize and identify other potential areas that need new or updated processes, policies, and procedures.

What This Person Is Expected to Deliver

Simply put, insight.

Because of the role's highly-skilled, multi-disciplinary nature, a Business Analyst's specific deliverables can differ from project to project. However, the role is most often expected to deliver business requirements documents when working on software development or implementation projects.

Another common deliverable a Business Analyst is expected to provide is *process flows* (think: boxes connected with arrows) for how work is currently done (*current state* or *as-is state*) and how work will be done after the software implementation or upgrade (*future state* or *to-be state*).

How This Person Approaches the Job

How Are Things Done Today?

"How things are done" is more formally known as *business process.* Business process is simply a set of steps an organization performs to accomplish its goals. In a perfect world, everything that an organization does is documented in a business process manual.

To ease the creation, maintenance, and understanding of business processes, they are often described at different levels of an organization. One example set of levels is

❖ Level 1: Organization

❖ Level 2: Department

❖ Level 3: Team

❖ Level 4: Individual

This is merely one example; there is no official, standard set of levels for defining business processes. You may sometimes see six levels or just three levels. You might see division or business unit above department. There is no one, right answer; it all depends on the how the organization is structured.

The important point is that there is not just a single process document that describes everything an organization does; there are multiple, and a best practice is to start documenting the process at the top of the organization and then decomposing it into subprocesses that the different organizational units perform.

Example: Large Oil and Gas Company

Level 1: The company runs its business and makes money by doing the following:

1. Exploring (for places where oil might be)

2. Extracting (oil once it finds some)

3. Refining (the oil into different products, such as gasoline for your car and jet fuel for airlines)

4. Marketing and selling (finding buyers for the products created in step three)

The next step would be to decompose each of these four steps into subprocesses. In other words, what are the steps the oil and gas company takes to explore? What are the steps for extracting? Again, in a perfect world, you would break this down until you have individual procedures that people perform as part of their jobs.

How Will Things Be Done Tomorrow?

Things change. Companies merge with other companies or expand into new territories. They develop new products and discontinue products that are not performing well. They close divisions when recessions hit. They implement new software to help with some portion of their business, such as time tracking or accounting. These changes can affect the way an organization does business—sometimes drastically—and what that really means is that people must do things differently or follow a different process.

It is usually the Business Analyst's job to define what the future state process should be by understanding what the process needs to accomplish, the tools or systems that need to be available, the people (departments, business units, etc.) involved in the future process, and any problems with the current state process that should be eliminated from the future state process. Then the Business Analyst iteratively identifies who needs to perform each step in the process and carefully lays it out.

Common Challenges

Developing and Maintaining Business and Technical Literacy

The Business Analyst is responsible for translating business needs into technical details. Therefore, the Business Analyst must be fluent in both languages, ensuring that nothing is lost (or worse, incorrectly described) in translation.

Being Expected to Represent the Business

Sitting squarely between the IT project team and the business stakeholders, the Business Analyst must often play the role of mediator between the two groups when issues arise. This means the Business Analyst is pulled back and forth between the two to help make sense of details. Also, because of the deep interaction with business stakeholders and the resulting insight into the stakeholders' needs, the Business Analyst is sometimes expected to speak on behalf of the stakeholders.

Being Asked to Wear Multiple Hats

Earlier I pointed out that the Business Analyst role itself requires skills in many areas. As a result, the Business Analyst is often called upon to support *any* effort that is dependent on the requirements captured for the project. For example, he might be asked to assist with writing test scripts or to help with developing training materials (especially if no formal organizational change management or training function exists in the organization).

Software Developer

What This Person Is Responsible For

Performing minor miracles.

What This Person Is Expected to Deliver

A Software Developer is expected to take a set of written rules or requirements, translate them into a mental image of a working solution, and then write code that builds this solution piece by piece. This mental image of a working solution is usually captured in a formal document known as a *Design Document* or *Technical Specification.*

The solution often takes the form of a tool or application that helps users do their jobs. The exact tool that the Software Developer provides varies from project to project, depending on the organization's needs.[7]

As mentioned earlier, one common tool the Software Developer provides is an application. Think about the apps on your mobile phone. Each app serves a distinct purpose. There are apps that allow you to buy and read books on your phone. There are other apps that allow you to edit pictures that you take with your camera.

Another type of tool may be an *interface* that moves data between

---

7. More information on this can be found in Chapter 10, "Components of an IT System.".

two systems. If you have ever downloaded your banking transactions from your bank's web portal into a spreadsheet on your computer, you have an idea of what an interface is. In this example, a Software Developer working for the bank wrote code that extracted the relevant transactions from the bank's database, copied them into a spreadsheet, and then allowed you to download them. The more common interfaces developed as part of an IT project follow steps similar to the following:

1. Extract data from the source system's database

2. Store the extracted data in a file

3. Move the file into an area that the target system can access

4. Load the data into the target system's database

How This Person Approaches the Job

To do their jobs, Software Developers need clear requirements. They generally build the simplest version of what they believe the users are asking for and then keep building on top of this, requirement by requirement.

Think of a Software Developer as creating a jigsaw puzzle, except she is not only putting the pieces together, she is designing the individual pieces themselves, and *how* they fit together. Each piece serves a purpose. A piece may represent calculations, a screen for users to interact with the software, or interfaces that move data from one system to another. In a complex system, the number of pieces can approach hundreds or thousands.

There are tools that the Software Developer uses to help keep track of these pieces, but a key skill is the ability to envision a model of the entire system (or specific pieces of it) that the Software Developer uses to understand how the various pieces fit together. To make this system model a reality, a Software Developer must consider the type of application to create, which we explore later in the "The Application" section in Chapter 10, and which programming language to use to create it.

There are many (many, many) programming languages, and these languages were generally created to address various needs. Just as a chef chooses different knives for different types of cutting or a mechanic uses different sets of tools when working on a car versus an airplane, the Software Developer must choose an appropriate language for the job. For example, there is a language called PERL (Practical Extraction Reporting Language) that offers Software Developers a Swiss Army knife of tools to manipulate text data, which makes it popular for building interfaces that require the ability to extract large batches of data, manipulate the data, and then transform it into a format that a target system can consume. C++ is a language that has been around for a long time, and specializes in performing calculations very quickly, making it popular in areas that rely on math, such as science and military applications. Python and PHP are two very common languages used for web application development. These are only some of the languages available for Software Developers to use when tackling a problem. Which language they ultimately choose is based on a combination of personal preferences, learning styles, types of solutions they need to provide, and of course, the needs of the organization they work for[8].

During application development, the Software Developer tests the application to see if it is functioning properly and if it is meeting the requirements. This is called *unit testing* and is akin to occasionally tasting a dish you are preparing as you add various ingredients and seasonings to it.

At some point, either when the Software Developer finishes implementing all requirements or as the project reaches certain milestones, the Software Developer releases code for other people to test (most likely the Quality Assurance (QA)/Testing Lead or Testers). If a Tester finds mistakes in the code (aka *bugs*, *defects*, or incorrect functionality),

---

8. Large organizations often standardize on a set of languages to ensure their teams have deep knowledge and capability in a handful of languages, as opposed to a broad knowledge in many.

then the Tester writes up this finding and sends it back to the Software Developer to be fixed. Otherwise, the application is ready for *user acceptance testing* (more on this in the "Quality Assurance/Testing Lead" chapter).

Common Challenges

Not Understood by Many

Because so few people who are not Software Developers understand what software development entails, it can be difficult for a Software Developer to effectively explain problems and issues that arise during development. It can feel like trying to explain calculus to someone who only understands basic math. Or perhaps a more accurate way of looking at it is like trying to explain a concept in Mandarin to an English-only speaker. The rules and underlying concepts are fundamentally different. This issue can arise when a Software Developer tries to explain to a Business Analyst or a user why a request cannot be easily implemented. More commonly this issue arises when a Software Developer must try to explain why developing a module (think: chunk of code) is going to take longer than expected.

How Long Will It Take?

The Software Developer must estimate how long it will take to build the solution, based primarily on the requirements gathered and little else. This can be an unsettling request—unless they've built this exact application many times before, Software Developers need to become comfortable with giving their best, most reasonable estimates to complete the development of their applications, based on the limited information they have available.

Having to Provide Solutions Without the Benefit of Context

Software Developers are expected to deliver fully functioning solutions from the set of requirements they are provided. However, they often do not participate in the requirements gathering sessions with users and so they may not have the benefit of context—perhaps resulting in solutions that meet the letter, but not the spirit, of the business requirements.

Staying Current

Many Software Developers must balance the demands of their day jobs with the need to expand their knowledge of new languages or new techniques. To be a good Software Developer is to be on a never-ending learning curve. Existing languages are constantly changing, and new languages are created all the time. Failing to stay on top of either means that a Software Developer's skills grow stale and eventually become obsolete (or more likely, only useful in an ever-shrinking market). This is often constrained by the needs of the job or the organization. If most applications that exist in an organization are written in only one or two languages, management may not see the need for Software Developers to broaden their skills.

> *Warning: opinion ahead!*
> *In my humble opinion, this is a mistake. Just as knowing multiple human languages expands the mind, offers new perspectives, and has numerous cognitive benefits, knowing multiple programming languages provides Software Developers with new ways of thinking about and addressing business problems.*

Quality Assurance/Testing Lead

What This Person Is Responsible For

The QA/Testing Lead is responsible for validating that the software being implemented works without error, and that it works according to how the specifications say it should work. The QA/Testing Lead determines the best ways to test the software product.

What This Person Is Expected to Deliver

A test plan that identifies what areas of the solution must be tested, as well as *test scripts* that are used to conduct testing. A test script is a step-by-step instruction explaining what a Tester should enter or do in a system and what the result should be.

The QA/Testing Lead is also expected to provide status about how the testing effort is going, usually by indicating which tests are complete and the percentage of passed/failed tests.

How This Person Approaches the Job

The best way the QA/Testing Lead can validate that a solution is working properly is to develop test scripts based on the requirements that the Business Analyst provided. Thoroughly testing the system should confirm that the solution successfully meets every requirement.

There are many different types of testing that the QA/Testing Lead organizes based on the type of project or stage of implementation. Some of the most common types of testing include the following:

❖ **Functional Testing**: Verifying that the application works according to the requirements. Testers—people who are trained in both software testing methods and the application itself—perform this testing.

❖ **User Acceptance Testing (UAT):** Testing that is similar to functional testing except business users conduct this testing to determine if the application meets their needs.

❖ **Regression Testing:** In the case of an upgrade or enhancement to a system, there is a risk that the recent changes may "break" another part of the system. For example, if Component A and Component B of a system both rely on a shared block of code, then changing this block of code to satisfy requirements to enhance Component A may cause Component B to malfunction. Regression testing is intended to catch instances such as this. By running the test scripts created for the application's previous version against the upgraded or enhanced application, the QA/Testing Lead can determine if the new enhancements broke any existing functionality.

These are the most common types of testing; however, there are many types, such as system testing, integration testing, usability testing, and stress testing.

Common Challenges

Being Dependent on the Timeframes of Others

The timing of the QA/Testing Lead's work is entirely dependent on completion of requirements, completion of development, and access to a stable environment in which to test. Additionally, testing is often one of the final stages an application goes through before go live. Those two factors put a great deal of pressure on the QA/Testing Lead to be ready to start exactly when expected.

Testing Against Incomplete Requirements

Testing can reveal deficiencies with the requirements themselves just as often as it reveals application defects. And unfortunately, sometimes the QA/Testing Lead does not have access to a comprehensive set of requirements. In these instances, the QA/Testing Lead needs to learn enough about the system to develop test scripts that cover the system's most important areas.

Cutover Planner

What This Person Is Responsible For

The Cutover Planner makes it all happen. This person is responsible for ensuring that all the steps needed to launch the new system (*cut over* to the new system) are done.

What This Person Is Expected to Deliver

A cutover schedule, which is an exhaustive list that lays out exactly what needs to happen and when everything needs to happen.

Returning to our earlier example of moving from one house to another, if you hired a Cutover Planner to handle the transition from your old house to your new house, you would receive a project schedule with a task list:

1. Moving truck arrives

2. Walk movers through home

3. Begin packing and loading

4. Cancel electricity account at old house

5. Cancel water account at old house

6. Cancel gas account at old house

7. Establish electricity service at new house

8. Establish water service at new house

9. Establish gas service at new house

10. Moving truck departs old house

11. Moving truck arrives at new house

12. Walk through new house

13. Begin unloading

and on and on. For each task, the list would indicate how many minutes it should take, the person responsible for executing it, and any links to other tasks. The overall plan would include additional details that may be helpful on the day of the move, like email addresses, phone numbers, and a map showing the route between the two houses.

How This Person Approaches the Job

Like a surgeon (queue Madonna beat).

Precision and detail is the name of this game. The person filling the Cutover Planner role is a highly detail-oriented person.

Because one of the major deliverables from this role is a very detailed project schedule, the Cutover Planner approaches this in much the same way as the Project Manager begins planning the project—the only difference is the goal. Whereas the goal for the initial project planning is to *build* a solution, the goal for the cutover planning is to *deploy* the solution.

The Cutover Planner works with each project team member (the Software Developers in particular) to understand what needs to happen as part of the cutover. The Cutover Planner also works with the IT Department to understand the *cutover window*, which is the timeframe during which cutover can occur—usually over a weekend. The Cutover Planner then creates a schedule that ensures all tasks can be done within this window.

Unfortunately, things don't always go according to plan (maybe some critical data couldn't be loaded in a key database for some reason). Therefore, the Cutover Planner also develops another schedule called the *rollback schedule*, which outlines what to do if the solution's deployment fails during the cutover window. The rollback schedule details the activities needed to "undo" the deployment.

Finally, to make sure that the actual cutover process occurs as smoothly as possible, the Cutover Planner arranges two or three rehearsals before the actual event. This is to ensure all participants know their roles and when to perform them.

Common Challenges

Uncovering Every Last Detail

Because building a cutover schedule demands painstaking detail, the biggest challenge is identifying and pulling together all the steps required to perform the cutover.

Coordinating a Massive Production

A cutover plan is similar to a play in that the success of the cutover depends on how well the plan is executed on opening weekend—everyone must know which part to play and recognize the cues. Therefore, the Cutover Planner needs to get everyone together to rehearse the cutover plan, direct multiple rehearsals, and make adjustments before validating the plan is thorough.

CHAPTER 8

Another Role You May Run Into

The Subject Matter Expert

What This Person Is Responsible For

When someone says, "I know a guy..." the "guy" is most likely a Subject Matter Expert.

The Subject Matter Expert (SME) is responsible for providing specialized knowledge and skills in a given area. For example, a project implementing SAP may have one or more SAP experts on hand.[9] Or if the project is improving processes in a hospital, there may be a member of that hospital working with the project team as a Subject Matter Expert, since that person is intimately familiar with how the hospital functions, what works, and what doesn't work.

Returning to the example of moving from one house to another, imagine that you have a piano that needs to be moved. Pianos are big, heavy, and delicate objects. Someone who doesn't know how to move one can easily damage it. For this reason, you might want to find a piano mover or at the very least have someone knowledgeable in how to move pianos supervise securing, loading, and unloading the piano. This

---

9. SAP is both the name of a software vendor and the name of one of its flagship products—an enterprise resource planning system.

person would be considered a Subject Matter Expert, with the subject matter being "piano moving."

What This Person Is Expected to Deliver

The Subject Matter Expert provides knowledge and expertise in a specific domain or topic that is relevant to the project. The "specific domain or topic" can be just about anything. It can be a functional area (such as accounting), a specialized skill (like moving pianos or writing code in C#.NET), or a topic (like the European financial markets).

For example, a project might need someone who is a Subject Matter Expert in the software being implemented to guide the team on the best way to configure that software for optimal performance. Or a project might have Subject Matter Experts for specific functional domains to clarify current state processes or to improve future state processes.

Because Subject Matter Experts are in fact experts, they can usually provide the project team with authoritative answers, best practices, and guidance more quickly than the team could learn on its own.

How This Person Approaches the Job

Like other project roles, Subject Matter Experts first begin by understanding what the project team is trying to accomplish. They then proactively look for opportunities to provide assistance—imparting knowledge and getting hands-on when appropriate. They may help the Business Analyst identify other requirements he may not have thought of before. They might also work with the Project Manager to identify ways to handle issues that arise.

Common Challenges

Balancing the Needs of Multiple Clients

Because Subject Matter Experts are, by definition, experts, they are often highly in demand by multiple groups. For example, a technical Subject Matter Expert for a software vendor or consulting company is usually spread thinly across multiple engagements. A functional Subject Matter Expert within an organization is usually the go-to person for just about everything in that function. As such, Subject Matter Experts usually have to juggle being involved in many initiatives.

Being Intermittently Involved

A consequence of being in demand by multiple groups is that a Subject Matter Expert's involvement is usually tightly controlled and short-lived. The Subject Matter Expert is expected to swoop in intermittently, quickly understand the full context and history of decisions and progress since the last swoop, and provide tailored and relevant advice to address the issue in question.

Selling Their Ideas Through Influence

The best practices and guidance a Subject Matter Expert gives is not always popular. The Subject Matter Expert may have advice for a client that will in fact solve a business problem; however, the client may dismiss this advice due to the culture of the organization or the level of difficulty involved in implementing the advice. This requires the Subject Matter Expert to then come up with solutions that take into account not only the business issue, but other components of the organization such as culture, available resources, and the level of change the organization is willing to tolerate.

CHAPTER 9

Stuff You Might Want (or Need) To Know

There Are Different Types of Project Managers

Project Managers are people too.

They have hopes and dreams, personal and professional challenges, and strengths and weaknesses just like all of us. While the same can be said of all team members, Project Managers are part of project leadership and, as leaders, they set the tone for the project team they lead—whether intentionally or unintentionally, consciously or unconsciously. Therefore, one determining factor of how well the project team functions and works together is the Project Manager's ability to lead.

Who's In Charge Is Not Always Clear

From time to time, you may notice disagreements about who is responsible for what or who is in charge of what. For example, some Project Managers view themselves as ultimately in charge of everything related to the project, such as the budget for the project and who they choose to bring onto or release from the project. Other people in the organization might view the Project Manager as the person who is simply there to execute their wishes. These people may believe that they should tell the Project Manager who is on the project team and how the project should be structured. In reality, both can work. There is no right way or wrong way. It depends on a variety of factors: the Project Manager's strengths and weakness, the project size, the current organizational culture, and ultimately who is picking up the tab.

Organizations Start Projects for Different Reasons

How do projects get created? What triggers an organization to initiate a project? The answers to these questions are as varied as the organizations themselves. In the next sections, I describe some of the more common reasons that organizations undertake projects.

Cost Cutting Measures

Companies start out small. Many operate in this state until they acquire so many customers that they must grow to meet the increased demand (think about a manufacturing company: more inventory, more factories, and more distribution points). This additional growth often requires more mature processes and systems to manage the additional infrastructure. If not managed carefully, this growth can result in redundancy and unnecessary costs. For example, an organization can have several systems that provide the same primary function but for different groups in the organization. The same is true of processes and this redundancy comes with a cost.

The cost of a system is not limited to its annual license fee or initial purchase cost. There are costs for simply maintaining a system. Organizations must purchase computer servers and hire personnel to maintain these servers (e.g., apply fixes and patches from the vendor, upgrade to the newest operating system, and protect the machine against viruses). Eventually, an organization may decide to reduce these costs by reducing the number of systems it maintains and streamlining the processes its employees follow. One example of this is replacing three accounting systems that are used in three different geographic regions with one accounting system. This allows the other accounting systems to be retired, saving on the license fees and maintenance costs associated with each.

Mergers and Acquisitions

Another scenario that results in many projects is a merger or acquisition. When one company buys another, it also buys all the existing systems and processes that are in the acquired company. If there is significant overlap in how the companies do business, this can result in significant redundancy. Therefore, one of the first tasks that leadership undertakes after an acquisition is to begin eliminating redundancy in personnel, in processes, and in systems. From an IT perspective, this means performing the same activities in the previous example ("Cost Cutting Measures"): identifying redundant systems, choosing which systems to standardize on, migrating everyone onto these systems, and then retiring the remaining ones.

Process Improvement

Another reason why a company might initiate a project is that the company simply realizes it is not being as efficient as it could be. For example, business stakeholders may say, "It should not take three weeks to process a simple invoice!" and decide to improve the process. As a first step, they most likely work with a Business Analyst who identifies problematic areas and comes up with suggestions to improve them. The Business Analyst might conclude, "It takes three weeks to process an invoice because you only have Mary in that department, it takes her an hour to process one invoice, and she gets a hundred a week. I recommend you either add more people to that department or implement an accounting system so that she can spend less time filing and sorting and more time approving and releasing payments."

Let's say the Business Analyst convinces management to change a process as part of a process improvement effort and management approves the proposed changes. How do you know things are actually better? Is it just because Mary looks less stressed? What if business slowed down and Mary could catch up for the first time ever? What if Mary simply gave up trying to keep up and said to herself, "I'm just going to process five invoices a day and to heck with the rest?" Does this mean things are actually better? No. In all examples, Mary could still be spending an hour on each invoice. So how do you know that things have improved? The only objective way to know for sure that a process has improved is by measuring.

Enter the world of *metrics*. Metrics are quantifiable, measurable units.

Let's conduct an example "look back" analysis to determine if a recent process improvement project met its goal.

Imagine you were working on a process improvement project for the Accounting Department of a large investment bank and that the Accounting Department is responsible for receiving, approving, and paying invoices the bank receives. The project concluded at the end of 2016. It is now 2017, one year after implementing the improvements, and you are trying to determine if the process improvements had any affect.

Imagine you are armed with the following data:

Metric	Before process improvements	After process improvements
Average # of invoices processed per month	20	36

Have things improved? Maybe. It is difficult to tell with this data. Was the increase due to efficiency gains or because the Accounting Department could begin catching up on a backlog of invoices? Maybe it also hired more people at the same time it implemented the changes and are now able to process more invoices as a result.

Challenge question: What additional data would you need at this point to determine if things have improved?[10]

Now, imagine that you have the following additional metrics for the same period to use in your analysis:

Metric	Before process improvements	After process improvements
Average # of invoices processed per month	20	36
Average # of invoices received per month	25	40
Average processing time per invoice	5 minutes	12 minutes

Have things improved? Here you begin to see how more data can provide additional insight. First, notice that the number of invoices received per month increased, which could be responsible for the increase in average number of invoices processed per month. So from a business perspective, it appears business may have picked up, since the bank is receiving more invoices (perhaps an indication that the bank required more outside help to handle an increase in demand). However, notice the last metric: the average time to process one invoice. It increased. If the goal of the project was to increase efficiency, then you might begin to suspect that it was not successful.

Finally, let's continue our analysis after receiving one more metric: the average time to approve one invoice.

---

10. To objectively determine if things have improved, you would need to see the data at a more granular level (e.g., month by month), instead of merely before and after. For example, having data for the 12 months leading up to the process change and for the 12 months after the process change lets you see whether the average number of invoices processed continuously rises, falls, or levels off, and thus begins to tell the story behind the data.

Metric	Before process improvements	After process improvements
Average # of invoices processed per month	20	36
Average # of invoices received per month	25	40
Average processing time per invoice	5 minutes	12 minutes
Average approval time per invoice	3 days	2 hours

Notice the significant decrease in time it now takes for getting invoices approved: from three days down to two hours! This most definitely represents a process improvement.

Here are the takeaways for any process improvement effort:

1. Get agreement on exactly what the organization is trying to improve. A Business Analyst works with the leadership and the business to define this and sets targets that determine if the project was successful or not.

2. Identify metrics to determine objectively whether improvement has occurred.

3. Use data to tell the findings. In our example, the amount of time spent processing each invoice increased from five to twelve minutes, while the average time to approve each invoice decreased from three days to two hours. This is all during a period when the average number of invoices received per month increased by 60 percent (from twenty-five to forty).

> **Note:**
> This is the type of clarity and insight that a good Business Analyst provides.

External Factors

Sometimes, factors outside of an organization's control require the organization to initiate a project. For example, commercial software vendors with large customer bases must carefully control the number software versions they actively support. Different versions of software inevitably have different sets of bugs and different ways of functioning. Customers are always demanding more features and better performance, which require code changes. Therefore, vendors must occasionally make tradeoff decisions between having developers spend time fixing old code or spend time writing code to provide new features and functionality. As a result, many software vendors set a cutoff for the number of previous versions they support. Any customers using software older than this cutoff are generally expected to make do with the software as is, bugs and all. Organizations often cannot afford to run software that is unsupported because the risk to the business is deemed too great. Therefore, to continue receiving vendor support and maintenance, organizations must upgrade their systems from time to time.

Other external factors that may require an organization to upgrade its existing software or switch to a new system are new laws or regulations where the organization operates. For example, many years ago the government implemented anti-money laundering regulations for banks as part of global crack-down on crime. One component of these regulations was the *Know Your Customer* set of processes, which specified the minimum amount of data that banks were required to collect about their customers so that the banks could positively identify their customers. If banks were not already collecting and storing this minimum data, they likely initiated projects to implement systems and processes to do so.

There Are Many Project Management Methodologies

There are many project management methodologies an organization can choose from when executing a project: Scrum, Six Sigma, Lean Six Sigma, Kanban, Rapid Application Development, PRINCE2, and waterfall are just some of the options. Then there are the internal, homegrown methodologies that consultancies and organizations develop. Each comes with a somewhat unique set of processes, terminology, and phases and each promises advantages over methodologies that have come before them.

The important thing to realize is that, at its core, each methodology proposes a way to break down the block of tasks that need to be done, a way to arrange and schedule each task for execution, and a way to measure progress.

Each methodology may also come with a set of roles. While the roles may not line up exactly with the roles described in this book, they often perform functions that are similar enough to them that you can understand their purpose.

Until recently, waterfall was one of the most popular approaches for project delivery. Now, many organizations favor Agile methodologies over waterfall as a way to execute projects.

Waterfall Versus Agile

Software development projects are usually complex initiatives. Because of this, over the years, people developed different approaches to understanding what users need and building systems that address those needs.

One approach that was popular in the 1980s and 1990s was the *waterfall* methodology, so called because the output of each step poured into the next step. Progress moved in one direction—starting with gathering requirements and ending with testing and delivering the software.

Large projects utilizing the waterfall methodology experienced a common problem: by the time the project delivered the solution, the solution was sometimes already obsolete. Here's how it would typically happen from the end users' perspective:

1. Someone would have a talk with us, asking us for details on some business pain point we were experiencing.

2. Then a team would grill us for several weeks or months, extracting every possible requirement from us in painstaking detail.

3. The same team would then send a giant spreadsheet or document with hundreds of requirements, asking us to "sign off" that the requirements were complete and accurate—or worse, the team would force us to sit through endless, torturous requirements review sessions to get the same sign off.

4. Once we signed off, the project team would disappear into a black hole, often for over a year.

5. One day—usually over a year later—we would receive an email saying that the new system would be live on some date, but first we needed to "test" the system, which usually meant executing scripts based on the endless requirements we blessed over a year ago ("Click here, then here, enter this, and click that. Do you see 150? Good, mark this as passed.").

6. Then we would be stuck with a system that may or may not actually provide us with what we need.

Now, there are *many* problems with this scenario. Here are two:

❖ For items two and three, very few end users think in terms of requirements. Users have more of a "I need to be able to do x" perspective, as opposed to typical, government-style requirements like "The system shall allow the user to enter price and quantity" and "The system shall compute order total by multiplying price times quantity." I'm not exaggerating here—the actual requirements are even drier than these examples and there are hundreds of them. Given this, it is difficult for an end user to look at several hundred lines of requirements written in something akin to Old Testament language and say, "Yes. This indeed meets all our needs." But faced with the threat of being the cause of delay in a multi-year, multi-million dollar project, the poor embattled end users eventually do sign off.

❖ For item four, once the end users signed off, they never really heard from anyone again. They weren't involved any further in the design. They couldn't provide any input on what was extremely important. But most significant, they couldn't alert the project team to changes in the organization, environment, or industry that might render certain requirements obsolete or necessitate new requirements. This resulted in software being delivered that was no longer fit for purpose.

In February of 2011, seventeen thought leaders in the software development space convened to discuss the state of how software was produced and delivered. From this discussion came a new philosophy known as "Agile." Agile is more of a mindset based around a set of principles and describes an ideal of how people should work together to develop solutions. There is even a manifesto and set of twelve principles on which the approach is based (these can be found at http://agilemanifesto.org.)

Following the introduction of Agile, a number of new delivery methodologies came on to the scene, each slightly different from the other, but that all embody the philosophy spelled out in Agile. One of the more popular approaches that fall under the Agile umbrella is "Scrum."

In Scrum, there are three primary roles.

First, there is a Product Owner who works with the end users to identify the work that needs to be done and the order (i.e., priority) of the work. The Product Owner functions as the main liaison between the development team and the end users.

Next is the Development Team who is responsible for doing the work necessary to deliver the product that the Product Owner envisions. This team is comprised of all the roles necessary to deliver the product, including Business Analysts, QA/Testing Leads, Technical Writers, etc.

Finally, there is the Scrum Master who has the dual responsibility of keeping the Development Team as productive as possible while also reminding everyone of the values that Scrum is based on. The Scrum Master serves as a liaison between the Product Owner and the Development Team.

One of the biggest differentiators between Scrum and waterfall is the timeframe in which work gets done. Whereas in waterfall, a project might be underway for a full year before anything of value (to the end user!) can be shown, Scrum aims to deliver demonstrable value to the end user as quickly and as routinely as possible.

Some characteristics of Scrum include:

❖ Instead of requirements, there are user stories. A user story includes the role the user is filling, a task the user needs to perform, and the user's goal for performing this task. Each story is written in plain English, from the user's perspective. An example user story is, "As an Order Entry Specialist, I need to enter price and quantity so that I can obtain the total price of the order."

❖ Instead of one, big-bang delivery, projects deliver solutions iteratively. They deliver functionality in fixed development windows known as sprints (usually lasting two weeks or one month each). For example, "We'll deliver a, b, and c functionality in sprint 1. Then x, y, and z in sprint 2."

❖ The first step in each sprint is to determine what will be delivered in the current sprint, with input from the Product Owner. This ensures that the team is focused only on the most relevant and useful functionality in each sprint.

❖ The Product Owner routinely engages with the end users in order to understand their current needs and business priorities. This gives users an opportunity to adjust priorities based on organization, environment, or industry changes that could make certain requirements obsolete. (In my opinion, this is the real strength of the Scrum methodology: the ability to quickly respond to changing requirements.)

This is but a brief description of Scrum. For more information, check out the online *Scrum Guides* at https://www.scrumguides.org.

CHAPTER 10

Components of an IT System

To help round out your knowledge of what's happening on your typical IT project, it is important to have an idea of what exactly an IT project team delivers.

In a construction project, it's clear that the project team is expected to deliver a building or a parking structure or some other edifice, as well as

- ❖ All the internal systems for the building (electrical, plumbing, ventilation)

- ❖ Architectural diagrams of the building, including how the various internal systems are connected to each other

- ❖ Various processes for maintaining and using the building (usually contained in an Operations Manual), such as how to contact the fire or police departments and how to shut off the water

The project team for a major surgery is responsible for operating on the patient, as well as

- ❖ Monitoring the patient for a period of time after the surgery

- ❖ Identifying the medication and activities (rest, light

exercise, no heavy lifting) needed to support the recovery process

❖ Updating all the patient's paper and electronic medical records

❖ Billing the patient and insurance company for the surgery

The point is that major projects—like IT projects that require teams to work and collaborate over the course of hours or months—deliver more than just the main thing that is being created. They also typically deliver additional components required to support and maintain the thing being created.

IT projects usually deliver some combination of the following set of components:

❖ A front-end application for users to enter, retrieve, or analyze data

❖ A database that stores the data

❖ The hardware and infrastructure required to host the application and database

❖ The processes that the users should follow to perform their job functions

With the exception of processes, which I describe in the "Business Analyst" section, I briefly discuss each of these components in the next few sections. While I do not cover all the latest trends and developments in IT hardware and software (the information would be obsolete by the time the book was published anyway), I do provide a foundation upon which you can begin to build your knowledge, if you choose to do so.

The Application

The application is what you use to get something done. When you post a picture of a dish from your favorite restaurant or view your latest banking transactions, you are using an application. An application typically consists of two main components:

❖ **A user interface,** which includes the screens or pages you look at when using an application. The text entry fields, the buttons you click to perform some action, and the colors form the user interface.

❖ **The programming code,** which controls what the system *does*. When you click a button labeled "Retrieve" to show your banking transactions, the application executes code that retrieves your transactions from the database and displays them in the user interface.

Types of Applications

Different business needs require different types of solutions. Sometimes the business needs to enter data into a system, have the system perform some processing, and then store the data in a database. This is one of the most common types of programming and is generally called *application development.* An ATM's software is example of an application (banking transactions are entered and stored). Another example is a college's student registration system that manages student details, majors, courses for each semester, and grades for each course.

Applications come in three main forms. There is the *desktop application* that runs directly on a computer or laptop. All the code required for the program is stored on the computer and when the application is running, it is running in the computer's memory (this is the Random Access Memory or RAM you may have heard about).

There is also the *web application.* Think about all the things you do online today: banking, purchasing books and groceries, shopping, sending or receiving money, and posting (hopefully) interesting stories to your social media. You use web applications to perform these activities and teams of Software Developers created those web applications. Web applications require a web browser to access and use them. Contrary to a desktop application, the code required for a web application is stored on a server.[11] When the program is running, it is running in the memory of the server and not your local computer.

Finally, with the rise of mobile phones and tablets, developers are creating *mobile applications* (usually referred to as *apps*) for these devices.

Interfaces

Sometimes the business needs to see or access data in one system from another system. For example, stock traders may need to see stock prices (held in various commercial systems) in their trading system or sales people may need to access email threads between their organization's customers and their Customer Service Department. These require interfaces to either copy or expose the data in the original system to the target system.

Another scenario where interface development is often required is for reporting applications. Some reports are complex and require data from multiple systems to provide the comprehensive view the report is designed to provide. A common practice is to pull all the required data from the multiple systems into a single database that is then used as the data source for the report. This is implemented by creating one or more interfaces per system that are responsible for pulling the data from the source system and storing it in specific tables of the database.

---

11. A server is a powerful computer that typically has large amounts of memory and significantly faster processing speeds to handle the demands of serving multiple users simultaneously.

The Relational Database

A database sits at the heart of nearly every enterprise-wide system. It replaces the old, paper file cabinets from yesteryear. By providing a centralized, electronic repository of critical data, a database offers many advantages over its paper-based counterpart. For one, it provides nearly instantaneous access to critical data for anyone in the organization, regardless of that person's geographic location. It also allows for new ways of aggregating, analyzing, and distributing data that the organization collects. Finally, it enables improved levels of organization resilience through new disaster recovery options. For example, if a paper-based organization loses all its files in a fire or flood, its ability to operate—and in some cases, even stay in business—is severely disrupted. If instead, the organization maintains its data in an electronic database, it can periodically back up the database; in the event of a disaster, the organization can be up and running (sometimes within a matter of hours) by restoring the database in a new location.

Before the Relational Database

In the olden days, one of the most common ways to store electronic data was in a single file called a *flat file*. For example, if a company needed to track customers and their vehicles, it kept this data together in one file. The file had a fixed format or layout that each row in the file followed. For example, a file for storing a customer and up to three vehicles would have the following format:

[CUSTOMER1][VEHICLE1][VEHICLE2][VEHICLE3]

Each time a customer added a new vehicle to their household, the company would first have to find and remove the old customer record, then create a new record with the customer details and the vehicle details.[12] One of the main drawbacks to storing data in flat files was wasted storage space. Even if the customer only had one vehicle, a certain amount of space for the second and third vehicles was still allocated, and storing data was expensive back then—hardware was not as cheap as it is now.

The other challenge with storing data this way was that it was very difficult to implement changes. For example, now imagine that the company wanted to track up to five vehicles per customer.

```
[CUSTOMER1][VEHICLE1][VEHICLE2][VEHICLE3][VEHICLE4]
[VEHICLE5]
```

Since the structure for the file changed, the company would need to apply the structure to all existing records. This meant that data effectively had to be "poured" into the new structure, which took a lot of time and processing. Manipulating data or the structure of the data was a difficult and cumbersome process that required specialists in this space.

A New Way

In 1970, a man named E. F. Codd came along with a better way of storing data. He called it the *relational database management system*. In this system, data is stored in a set of tables that are linked to each other by *keys*. Each table contains or represents one type of object (e.g., a CUSTOMER) and each row in that table is represented by a unique key (e.g., a Customer ID).

One way to think of a relational database is like a spreadsheet workbook with different tabs. Each tab represents a table in the relational database. A database table has rows and columns, just like a spreadsheet does. Think of a spreadsheet table with headers. Those headers repre-

---

12. There were other approaches for updating data, and each had its pros and cons. For example, in addition to the customer and vehicle fields, a date field could be added that identified the date the record was created. The program would then need to find the customer record with the most recent date to find the customer's vehicle information.

sent the column names in the table.

Returning to our example of customers and vehicles, in the relational model there is one table for customers and another table for vehicles. The two tables are linked by the Customer ID, since vehicles belong to a customer. In this way, it is easy to retrieve all vehicles for a single customer simply by searching for vehicles with a Customer ID that matches the specified Customer ID. Refer to the rows in the following tables where the Customer ID is 1; this is how Sandra Petersen and her two vehicles are stored in this relational model.

Customer Table

Customer ID	First Name	Last Name
1	Sandra	Petersen
2	Tim	Johnson

Vehicle Table

Vehicle ID	Customer ID	Year	Make	Model
1	1	2008	Ford	Taurus
2	1	2011	Ford	Explorer

Notice the advantages of this approach over the flat file approach. First, no space is wasted. If there is no third vehicle, no additional space is taken up. Second, this design allows for an unlimited number of vehicles for each customer. If a customer acquires another vehicle, all you need to do is add a new row to the VEHICLE table.

Notice also that linking the tables together by the Customer ID column means you cannot add any vehicles to the VEHICLE table unless the Customer ID exists in the CUSTOMER table. This is to ensure that there aren't any vehicles in the VEHICLE table for customers that do not actually exist. This is the biggest strength of a relational database: data integrity.

Adding another table to the database requires defining the structure of the new table, and then the relationship(s) with existing tables. Let's say we now want to track vehicle maintenance history. We have a table to store customers, and another table to store the vehicles that belong to each customer. Now we need a table to store the various types of maintenance that are performed on the vehicles in the database. The structure of this new table will contain fields for capturing the type of maintenance activity (oil change, tire rotation, etc.) and the date the maintenance was performed.

For relationships, the new table will be linked to the VEHICLE table, since vehicle maintenance is a characteristic of a vehicle, and not the customer.

Notice that the existing data does not need to be touched. Instead, we are just adding a new structure to capture the new data. This is another strength of the relational database – being able to quickly and easily extend it without having to touch other areas of the database.

In our example we have three simple tables. Databases that support complex systems often have hundreds. It can be very difficult to keep track of all of the tables, their fields, and the relationships between them without some sort of visual aid. To help with this, a data model is often created. You can think of a data model as one of those directories you'll find in a mall (but without the "You are here" indicator.)

Let us examine the data model of our current database.

Note that the model contains the tables in the database, the data

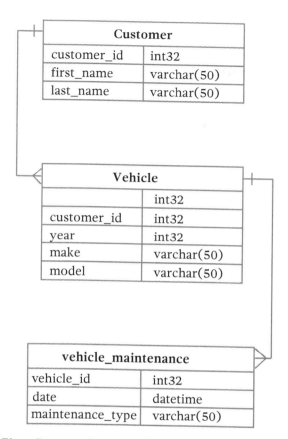

Fig 4. Data model of Customer Vehicle database

fields for each table, and the relationship between each table.

Looking at our model, we can see that one customer can have multiple vehicles (this is called a "one to many" relationship) and that each vehicle in turn can have multiple vehicle maintenance log entries.

Conclusion

Just about every website or app you use relies on a database of some sort. Anytime you post to your social media account, it gets stored in a database table. Anytime you read someone else's post, it was retrieved from a database table. When you registered for classes or received your grades, a database was involved. The database can be considered long-term memory of an application.

So, the next time you laugh out loud over a meme your friend posts on her feed, thank E. F. Codd, father of the relational database.

Infrastructure

The prefix *infra-* is of Latin origin and means *below*. Therefore, *infra-structure* is the structure below the system and it provides or supports the system's needs.

Examples of Infrastructure

A banana bread recipe by itself is merely a set of steps until someone comes along with a mixer, a pan, and an oven and executes the instructions—thus producing a loaf of banana bread. Similarly, a computer program is merely a set of instructions that requires a computer processor, hardware components, and other resources to perform its intended function.

The computer processor is responsible for reading the program steps and executing the corresponding actions (e.g., retrieve this data, calculate this result, display that value in this text box). The hardware includes components such as a monitor, mouse, and keyboard. Some of these components, such as its wired or wireless networking devices, enable the computer to connect to other computers, allowing it and programs running on it to access resources such as databases or other applications.

Another example to illustrate infrastructure is a house. Most people in the United States would not consider a house to be habitable until a number of connections are made to it. For example, it must have a way of receiving water, distributing water throughout the house, and removing waste water from the house. The same can be said of electricity. The house must be connected to the city's electrical grid before it can distribute electricity throughout the house. Thus, the infrastructure required for the house consists of water pipes and electrical wires provided by the city, in addition to the internal piping and wires that deliver each throughout the house.

Collectively, all the hardware and networking components required for a system to run is known as *infrastructure*.

The term infrastructure is not unique to information technology. The definition of *infrastructure* from the Oxford Dictionaries[13] is as follows:

> *"The basic physical and organizational structures and facilities (e.g., buildings, roads, power supplies) needed for the operation of a society or enterprise."*

Adapting the definition for an IT context yields the following definition for *application infrastructure*:

> *The basic hardware and network structures and components (e.g., servers, networking routers) needed for the operation of a system or application.*

For example, an application may require the following infrastructure:

❖   One server to host the application code

❖   One server to host the database

❖   Network hardware and cabling to connect the two servers together and allow users' computers to connect to the server hosting the application

Infrastructure Maintenance

As technology improves and applications require more and more

---

13. "infrastructure," *Oxford Dictionaries*, https://en.oxforddictionaries.com/definition/infrastructure.

resources to function, an organization may find that its existing infrastructure can no longer meet the needs of its operations. Much like a city whose freeway capacity can no longer support a significant increase in population, an organization may undertake one or more infrastructure improvement projects to upgrade its infrastructure to meet existing and anticipated needs. The organization may choose to include this improvement effort in the project scope. Or in some cases, the organization may realize it needs major infrastructure improvements in many areas and initiates an infrastructure improvement project to meet the technical requirements of multiple projects.

What an IT Project Typically Delivers

So, what does this all look like in the context of an IT project? What is the project team expected to deliver by the close of the project? The answer to this depends on what the organization needs and what it might already have in terms of systems, infrastructure, and business processes. Let's follow an organization through three stages of system usage. For each stage, we look at the type of project an organization typically undertakes and the impact each can have on IT system components.

Assume an organization still uses paper and pen for everything and finally decides that using a computer system will make it more efficient. Further assume that the organization already selected a system to use. In this scenario, the organization decides to undertake a *system implementation* project in which the project is responsible for installing a system and moving the organization's business from one system (pen and paper) to another (the selected electronic system). Here are some of the tasks the project team is expected to perform (not necessarily in this order):[14]

❖ Work with the organization's leadership to understand the goals of the project.

---

14. This is an extreme simplification of what actually happens in a system implementation project. There are entire books dedicated to these types of projects.

❖ Meet with business users to understand their processes for doing their jobs.

❖ Configure the system so that it matches the business's processes as much as possible.

❖ Work with the business on processes that need to change to accommodate the system.

❖ Purchase and install all computers that the users require.

❖ Purchase, install, and configure the servers and necessary infrastructure to connect everything together.

❖ Train users how to use the system to do their jobs.

❖ Test and retest the application to make sure it is working as intended.

❖ Facilitate user acceptance testing.

❖ Deploy the software to each user's computer.

❖ Provide support to users having problems with the new application.

As you can see in this scenario, the team is expected to deliver and install all components of the system, including the application, database, and all required infrastructure.

Continuing the example, let us now assume that the organization has been using the application for a number of years and the vendor announced that an upgrade is available. After reviewing the list of changes in the new version, the organization's leadership determines that it makes financial sense to switch to the new version. In this scenario, the business expects the project team to deliver a new version of the application. This is known as a *system upgrade* project and follows many of the same steps as the initial system implementation, as well as a few new ones:

❖ Evaluate the new version's infrastructure requirements, compare them with the organization's existing infrastructure, and make recommendations on which infrastructure components, if any, need to be upgraded.

❖ Set up temporary infrastructure (sometimes called a *test environment*) where the project team installs, configures, and tests the new version of the application.

❖ Over a weekend, execute the cutover plan to switch the organization to the new version. Ensure that all data is migrated from the old version to the new version correctly.

The vendor most likely made changes to the database to support the application changes, so the project team may need to upgrade the database as part of the cutover process too. Furthermore, the application's new functionality may require the project team to review and possibly modify business processes to accommodate it.

Continuing the example, let us now assume that after many years of using the system, leadership decides that the system no longer meets its needs and that it needs to replace it with an entirely new system. In this scenario, the project team performs all the steps from both the initial system implementation and the system upgrade, starting with the following major activities:

❖ The project completely replaces the application.

❖ The project may need to upgrade or replace infrastructure, depending on the new system's technical requirements.

❖ As with the upgrade, the project will need to review all processes to determine how the new application can accommodate them.

CHAPTER 11

Putting It All Together: a Mock Project

The Setup

In this chapter, I use a mock project to illustrate how project roles work together to help the business users do their jobs more effectively. I deliberately chose a simple (even silly and impractical) objective as the primary goal of this project: enter invoice amounts and receive a total amount. I chose this example because I want you to understand the interplay among the roles—not the subject matter the project is addressing. So, as you read through this story, if you find yourself asking questions like "Why don't they just use a spreadsheet?" or "Doesn't their accounting software do that?" you are right to do so. Just pretend that, for whatever reason, this is the selected option. Instead, focus on what the various team members on the project are doing. Try to predict each decision's outcome before continuing to the next section. Finally, I organized the story by project phase so that you can see the different actions each role takes in the various phases.

Background

The Accounting Department at XYZ Corporation (XYZ) needs a faster way to obtain the total amount for all invoices it receives on a given day. Today, some accountants use paper and pen and manually do the math to get the amount. Others use a calculator, but this causes problems when they realize they entered a number incorrectly and must go back and reenter all the values again from the beginning. The department is looking for a solution it can use to enter a list of invoice amounts, click a button, and get the total amount.

The Accounting Department manager finally decides to address this issue. She contacts the head of IT and explains the situation and what she would like. The head of IT, in turn, commissions the project and appoints a Project Manager.

The Invoice Project Starts

Initiating: The administrative phase. Creating the project charter (something like a birth certificate for projects), identifying the objectives for the project, and securing funding.

The Project Manager is hired and starts two weeks later. The head of IT briefs the Project Manager on why he commissioned the project, the goals of the project, and the Project Sponsor's name and title. They agree that the Project Manager is responsible for hiring the rest of the team needed for the project and for the overall project budget.

Knowing that a good relationship with the Project Sponsor is important, the Project Manager gets started by introducing himself to the Accounting Department manager. He discusses the importance of the project with her and how she expects the project to help her department. They also discuss the frequency of status updates and if she prefers status updates by phone or in person.

Armed with this initial information, the Project Manager gets to work on building the team and completing any necessary paperwork that XYZ mandates. He decides to bring a Business Analyst onto the project first, to start gathering requirements. The Business Analyst gathers knowledge that also helps provide key information needed for planning.

The Team Plans the Project

Planning: Thinking holistically about what needs to happen from beginning to end, breaking it down into chunks or major blocks of work, and then breaking down these chunks even further to the task level. Planning also includes proactively thinking about what could go wrong and putting a plan in place ahead of time (aka Risk Management Plan).

The Project Manager begins to plan the project. He decides to create an initial, high level schedule, identifying all the major tasks and subtasks that the project needs to complete. Based on his years of experience with these types of projects, he expects that the project will be comprised of the following major tasks:

    Task 1: Analysis

    Task 2: Software development

    Task 3: Testing

    Task 4: Deployment

With these major tasks identified, he begins to break down each major task and estimates how long each subtask will take. For Analysis, he identifies the following subtasks:

    Task 1.1: Stakeholder interviews, 5 days

    Task 1.2: Requirements development, 5 days

For Software development, he identifies the following:

    Task 2.1: User interface development, TBD

    Task 2.2: Calculation code development, TBD

```
Task 2.3: Unit testing, TBD

Task 2.4: Documentation, TBD

Task 2.5: Transition (all tasks related to handing the
application over to XYZ's IT Department)
```

Because he is not a Software Developer, the Project Manager does not know whether this is considered a complex or an easy application to build, and therefore decides to rely on the Software Developer to provide the estimates. He then finishes identifying the subtasks for the remaining major tasks. Finally, after breaking down all tasks into subtasks, he schedules a meeting with the team to review the schedule. Before the meeting, he sends the draft schedule to the team to give them time to review it and come up with feedback.

In the schedule review meeting, the Project Manager asks each team member to confirm that all the necessary tasks are identified and that the initial estimates are reasonable. He also gives each person an opportunity to revise the estimates if needed.

After some discussion and modifications to the schedule, the team agrees that it is a good schedule. The team estimates that the overall project will take four weeks—approximately one week for each major task.

Satisfied that he has a workable plan and schedule, the Project Manager organizes a planning review session with the leadership. In the session, he walks the Accounting Department manager and the IT manager through the project schedule. He also reviews the risks he identified and the mitigation plan for each. The leadership is satisfied with the plan and schedule and gives the OK to proceed.

Project Execution Commences

Executing: Where the rubber meets the road. Pulling the trigger and commencing work. During this phase, the Project Manager "simply" ensures that there is nothing standing in the way of starting and completing tasks. In IT projects, the Executing phase often has specific work streams for Requirements or Analysis, Design, Development, Testing, and Go Live.

The Business Analyst Gathers Requirements

During the first week (as planned and identified in the project schedule), the Business Analyst conducts a series of meetings with a small, representative group of Accounting Department members. She asks them questions about their jobs to understand the environment in which they operate and the context in which they will use the application.

Next, she begins to discuss details regarding the application, including how they will use it, who else might need to use it, and other systems it may need to connect to. She finds out that the accountants need to be able to enter up to twenty invoices in the system (they never receive more than that) and that it should simply add up the invoice amounts and give the accountants a total amount. The accountants store this figure in another system each day and this other system generates an annual revenue report. The Business Analyst makes her conclusions and then writes the following requirements:

REQUIREMENT #1: The system must allow the user to enter twenty invoices.

REQUIREMENT #2: The system will calculate the total amount for the invoices once they have been entered and will display this total on screen.

Satisfied that she captured the necessary requirements, she emails these to the Software Developer. She copies the QA/Testing Lead on the email because she knows he is looking for the requirements to build his test scripts.

The Software Developer Builds the Initial Application

The Software Developer receives the requirements a day before the date specified on the schedule. She is pleased with this because it gives her a little more time to study the requirements and design an application to meet them.

After reading the requirements, she decides to design an application that will have twenty textboxes labeled #1, #2, and so on. At the bottom of the screen, there will be a button with the words "Calculate Total" on it. When users click this button, the application will check that all twenty textboxes have amounts entered and then will display the total amount in a pop-up window. She names the application "Invoice Totaller v1.0" and gets to work.

She writes the code in C#.NET because that is XYZ's standard programming language for applications that run on users' desktop computers. After around five hours of designing the user interface, writing the code, and conducting unit tests, she believes it is ready for the next level of testing. She checks her code into the repository and sends an email to the QA/Testing Lead (courtesy copying the Business Analyst) to notify him that v1.0 is ready for testing.

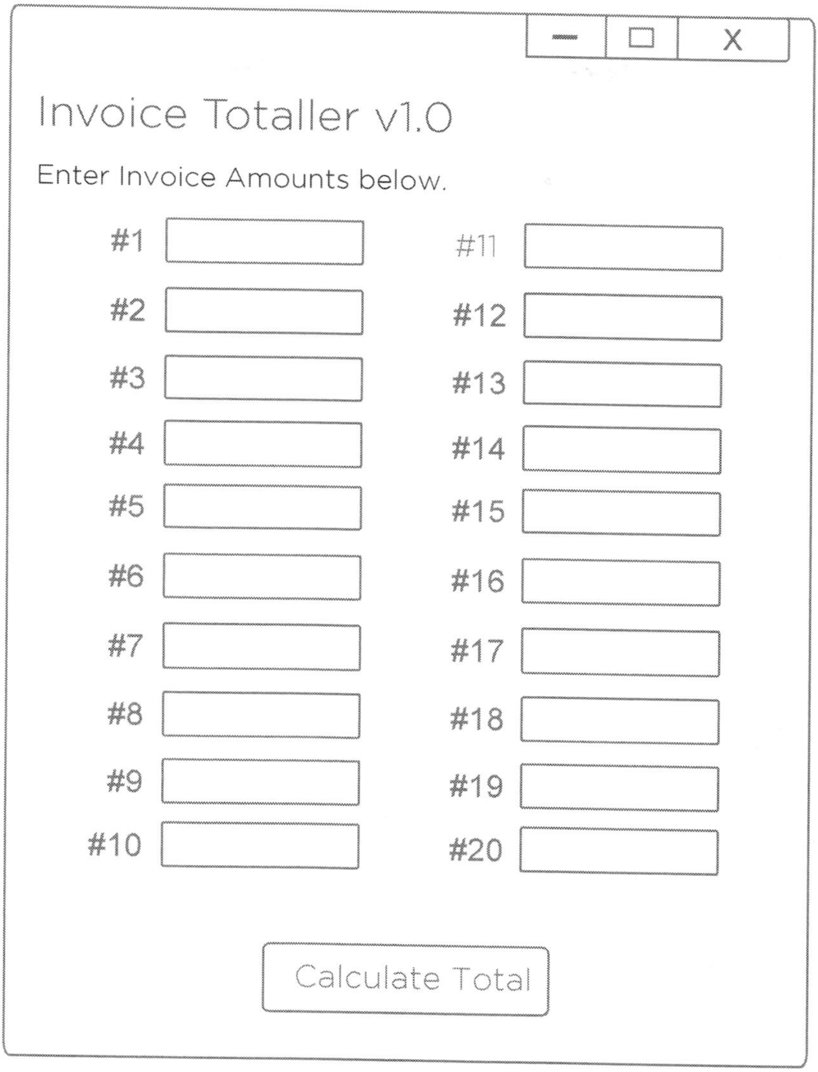

Fig 5. Invoice Totaller user interface.

The QA/Testing Lead Tests the Application

After the QA/Testing Lead receives the email from the Software Developer informing him that the code is ready to be tested, he begins to test the code against the test scripts he created based on the initial set of requirements received from the Business Analyst.

First, the QA/Testing Lead tests to see if the application is functioning according to the requirements. He enters twenty randomly selected amounts and then clicks the Calculate Total button. The application displays the expected number.

Next, he only enters five numbers and then clicks the Calculate Total button. The application responds with a message box indicating that twenty invoices need to be entered before it can calculate the total. He enters zeros for the remaining fifteen invoice amounts and clicks the button again; it responds with the correct total.

With this complete, he is satisfied that the application is working as designed, and that it is designed according to the requirements.

Once he completes his functional testing, he considers whether the users should test the application, which is what XYZ's methodology for deploying IT applications specifies. He decides that the application is simple enough that it does not warrant the extra effort needed to arrange UAT sessions. Instead, he signs off that the application is ready for deployment.

The next day in the team status meeting, he reports this as his status and explains his rationale.

The Users Give Feedback

The Project Manager coordinates the handover of the application from the project team to the IT Department; the handover also includes packaging all the documentation the project generated and storing it in the IT file repository. The Software Developer conducts a code walkthrough with the IT Department's Software Developers. Finally, the project team trains the IT Department's Business Analysts and Help Desk personnel how to use the application.

When the IT Department is comfortable that it has everything it needs to deploy and support the application on its own (without the project team), it signs off on the transition. It deploys the application the following Friday after the close of business, which gives it time to package the application for deployment to the Accounting Department computers. The IT Department selected this Friday evening timeslot to give it time to address any problems that might arise during the deployment—without interrupting business.

On Monday morning, the Accounting Department starts to use Invoice Totaller v1.0. Almost immediately, the users begin calling the Help Desk with issues, the most common being that they get an error message whenever they enter fewer than twenty invoice amounts. The Help Desk tells them to enter zeros for the remaining amounts (as the Help Desk learned during the handover training).

The number of calls to the Help Desk eventually decreases as word gets around about entering zeros when there are fewer than twenty invoices. But the Accounting Department members assert this is causing more work for them since previously, they only needed to work with whatever amounts they had.

The Accounting Department members eventually go to their manager and say that the application is causing more work than before, and that they believe they are getting less work done. Realizing that her team's productivity is suffering, the Accounting Department manager has a word with the IT manager to explain the situation.

The Team Modifies the Application

The Business Analyst meets with representatives from the Accounting Department again to discuss the issues. She learns that her original requirements did not accurately reflect the department's needs. While it's true the users need to enter up to twenty invoice amounts, the department often receives fewer than five invoices per day.

The Business Analyst adds a third requirement:

> REQUIREMENT #3: When no amount is entered for an invoice, the system should automatically treat that as a zero value.

The Business Analyst then sits down with the Software Developer to make sure the Software Developer understands how the application will be used. She walks the Software Developer through how she sees the application.

Once the Software Developer understands what the Business Analyst envisions, the Software Developer begins to modify the application. She adds the necessary code to automatically substitute zero for any amounts not entered, and then begins to conduct her unit tests.

First, she enters one invoice and clicks the button to test if her changes are working. The application correctly shows her the value of the one invoice and it does not display an error about the blank invoice amounts this time. Then she enters five invoices and clicks the button. Again, the application correctly displays the total of the five invoices. She conducts two more tests: entering zeros in all twenty fields and entering actual invoice amounts in all twenty fields. Both times the application displays the expected amounts.

Satisfied that the application is working according to the new requirements, she notifies the team that the code is ready for further testing.

The Project Consults the Change Manager

Up until this point, the Change Manager had not been involved with the project. After witnessing how the v1.0 rollout went, the Project Manager approaches the Change Manager and asks for her assistance with the rollout of v2.0. The Change Manager has seen other project teams mishandle change before and so she is not surprised. She knows that not having UAT means that the users won't see the application until after it is already deployed and on their desktops. She also knows from her involvement with previous projects that the Accounting Department staff generally has different ways of calculating invoice totals—some use paper and pencil, others use spreadsheets. She knows that it is very unlikely that the project can build an application that addresses the entire team's needs without involving any team members in the application's design and testing.

She decides that the best way she can support the rollout of v2.0 is with a combination of communications and one or two engagements. She also suggests that the project team plan on sitting with the Accounting Department for one or two weeks immediately after go live to answer any questions and quickly address any issues that may arise. As she thinks it through, she writes down her deliverables and ends up with the following communication and engagement plan:

- ❖ One week before go live, a "Heads-up" email, alerting users that v2.0 of the application is scheduled for deployment

- ❖ The Friday before go live, a "Reminder" email, letting users know that the application will be available on Monday morning when they come in for work and how to get help

- ❖ Early Monday morning, a "v2.0 is live!" email, letting users know that the application was successfully deployed and is available for use

- ❖ A "Post-go-live support ending" email, letting users know that the project team will be leaving the floor and that the users should return to calling the Help Desk for support

- ❖ A lunch and learn engagement that users can attend if they are interested in seeing the new version of the application and if they have any questions or concerns

❖ A cheat sheet for the UAT participants

❖ A post-go-live support model, detailing which team members will sit in the Accounting Department and how long they will provide this enhanced support

She meets with the Project Manager and the Business Analyst to share her recommendations and they agree. The Business Analyst suggests that they invite the Accounting Department manager to the lunch and learn engagement, since the manager's attendance would be seen as a sign of support and encourage other team members to attend. The Business Analyst also suggests that the cheat sheet for the UAT participants could be useful for the remainder of the Accounting Department. The Change Manager thinks these are both excellent ideas and makes a note to add the manager to the invite list and to leave a copy of the cheat sheet on everyone's desk before going home on Friday evening.

The Change Manager then meets with the Accounting Department manager to explain the plan. The Change Manager asks the Accounting Department manager if she would be willing to attend the lunch and learn and to say a few words at the beginning of the session to kick things off. The Accounting Department manager agrees to do so despite her busy schedule. The Change Manager completes the necessary logistics: sending the meeting invitations (including one to the Accounting Department manager), booking the room, ordering lunch, and assisting with putting together the presentation.

The Change Manager then meets with a couple of Accounting Department members to share a draft version of her cheat sheet and get some feedback. They point out a few areas that could be clearer and the Change Manager incorporates their suggestions.

Finally, The Change Manager writes the emails that the Project Manager will send as part of the communication plan. She reviews the drafts with the Project Manager and the Business Analyst. When finalized, she packages the emails in a document and provides the document to the Project Manager so that he can send the emails on the correct dates.

To ensure the emails are sent as planned, she sets calendar reminders for both herself and the Project Manager, reminding them when to send each email.

The day of the lunch and learn arrives. To start the meeting, the Accounting Department manager talks for a few minutes about how the application will be helpful to the department not only today but also in the future as new systems come online. The total invoice amount that the system calculates will eventually feed automatically into the new systems, reducing data entry work the team is currently doing. She then turns the floor over to the Business Analyst who provides a walkthrough of Invoice Totaller v2.0. Since the Software Developer is still modifying the application, the Business Analyst shows mockups of the system. She also shares tips and tricks like keyboard shortcuts for quickly moving from field to field. And of course she makes a point to bring up that the users no longer need to enter values for all twenty invoices before being able to get the total. She then solicits the group for any feedback and input. The Accounting group seems happy with the changes. The users have a few questions but are otherwise satisfied with the improvements.

The QA/Testing Lead Retests the Application

This time, the QA/Testing Lead decides to conduct all the testing that XYZ's methodology specifies. First, he completes his functional testing by performing the same type of testing as the Software Developer. To ensure that the changes did not break the application, he then conducts regression testing by running his original test scripts against the application.

Finally, he organizes several UAT sessions. He invites four Accounting Department members to run through the test he designed. The users test the application, leveraging the cheat sheet that instructs them how to use the system.

The users can see the changes right away. They are happy that the application is simpler, and that they do not have to enter twenty values before being able to obtain a total. After running through the process several times on their own, they are satisfied that the new application will work for their department. They sign off, which indicates that they accept the system as it is currently functioning, and the QA/Testing Lead notifies the project team, including the Project Manager.

The users have additional suggestions for making the application even better, which the Business Analyst captures as possible enhancements in future versions.

The Project Delivers the Application

The Project Manager begins sending the communications that the Change Manager prepared for him, starting with the heads-up email one week before go live.

The IT team deploys Invoice Totaller v2.0 without a problem the following Friday, after the close of business.

On Monday, the Accounting Department users begin using v2.0 of the application and are pleased with the changes. As planned, several project team members are located on the same floor as the Accounting Department so that users can walk over to ask questions as needed.

By the middle of the second week, the project team only has two requests for assistance. At the end of the second week, the IT Department is confident it can take over support for the application and gives the Project Manager permission to close the project. The Project Manager sends a final communication to the Accounting Department members, letting them know that support for Invoice Totaller v2.0 is transitioning back to the Help Desk as their first point of contact.

The Project Manager Monitors the Project

Monitoring and Controlling: The Project Manager tracks progress, monitors for risks (and triggers the Risk Management Plan if needed), and updates the schedule with status and new tasks that have been identified; the Project Manager constantly assesses how these things impact the completion date. This phase happens throughout the entire project.

Given that it is a short project and expected to last no more than four weeks at the most, the Project Manager establishes a daily checkpoint meeting. This meeting is scheduled for 4:00 p.m. to 5:00 p.m. and the entire team is expected to attend, so that everyone can know the status of the project and begin making any required plans.

The Project Closes

Closing: The wind-down phase. In IT projects, you do things like conduct lessons learned sessions (what went right, what went wrong) and release resources. Several months after go live, you may also evaluate success metrics to determine if the project delivered the intended benefits.

Thirty days after go live, the team does a final review of the number of Help Desk tickets submitted for the new invoice system. The team determines that very few have been submitted, and the need for enhanced support is over. The project team disbands and it hands over the new invoice system to the Help Desk as primary support.

The Project Manager completes final documentation according to XYZ's policies. He has the Project Sponsor sign off that the project was delivered successfully. He sends a message to the entire team, thanking the team members for their hard work. He also sends personal messages to the team members' various managers, letting the managers know how well the team members performed.

One month after the project team disbands, a user from the Accounting Department calls the Help Desk. The user tells the Help Desk agent that he has an invoice amount that he can't enter in the system because the number is too large to fit in the field provided by the application...

Lessons From the Invoice Project

❖ Requirements must be clear and unambiguous.

❖ Involve users early and often. Review requirements with them, review the design with them, review the support plan with them, review everything with them when possible. They sometimes provide context or ideas you may not have thought of.

❖ The steps and processes included in methodologies are generally there for good reason. If you decide to deviate from the guidance the methodology provides, understand the risk and get feedback from other stakeholders.

❖ Involve the Change Manager as soon as possible—ideally when the project is being conceived and preferably not after planning is complete.

❖ By not involving the users in the application's design and initial testing, the project took longer and cost more than originally planned.

❖ Applications grow and change over time. Users will always have suggestions for improvements and there will always be bugs in the code or scenarios that the application cannot handle. This is normal. It never ends.

CHAPTER 12

Tips for Succeeding at Your First IT Project

(and Leveraging the Experience to Boost Your Career)

B efore you go, I'd like to share some suggestions with you. These are tips I often share with people entering the world of IT projects.

Tip #1: Embrace this new opportunity

With so many people around who all speak a language (or languages) that you don't understand, it is easy to be intimidated and begin to think these people are significantly more intelligent than you. This is not the case. They are simply more experienced than you. Remember, if you weren't intelligent, then it is highly doubtful that you'd be here.

It's not wrong or even unexpected to feel this way. In fact, so many have felt this way that this feeling has its own name: *Impostor Syndrome*, which is the feeling that you don't belong here and it's only a matter of time before someone finds out.

While we're on the subject of intelligence, understand that it's not arrogant to remember that you're intelligent. I consider intelligence a measure of how well you can learn, as opposed to how much you know. Look at it from this angle: you've learned enough to get here. After all, if you're on a team full of geniuses, someone must think you belong there! Realize that and hold on to it. By reading this book, you are doing exactly the right thing to get the information you need to succeed in this space. Before you know it, you'll be waxing lyrical about decision support packages, phase gates, and go/no go checkpoints.

Takeaway: Accept that feeling intimidated is par for the course. Just hang in there. Take it one day at a time, learning one new thing each day or week until the day comes when your experience catches up with your intelligence and you begin to feel like you know what's happening and your role in all of it.

Tip #2: Focus on helping the project team succeed as a whole, not just on performing well in your own role

In our society, we've been raised to believe that as long as we do our best, that is enough. Or that we only need to worry about our own performance, our own grades, or our own success. So, it's reasonable to feel the same way on the job: as long as I worry about my own project deliverables—my own individual contribution—I don't need to worry about what other team members are doing.

This is wrong. Not only is it wrong, it can be harmful.[15] When you focus purely on your contributions without considering the needs of your fellow team members, you can cause rework, overtime, schedule delays, and cost overruns. Also, being concerned about your team members' wellness, needs, and performance is not just touchy-feely—it is good for your personal bottom line. Being the type of teammate that helps your team perform at its best is good for your reputation, which helps you earn recommendations and offers for additional work.

Takeaway: Learn how to better address your team members' needs. As you meet with various users and project team members, identify ways to provide them with what they need.

---

15. This is a personal philosophy of mine reinforced during my time in the Marine Corps. As a Sergeant, I was graded not only on how well I looked after my own fitness and performance, but also that of the Marines that reported to me.

Tip #3: Leverage your experience on the IT project to boost your career trajectory

The time you spend on an IT project can yield a number of measurable benefits, including:

❖ Knowledge about project management, a skill that is useful in a number of industries, technical and non-technical alike

❖ A greater understanding of how the overall business functions from engaging with other departments

❖ A better corporate network from the various interactions with the people you'll work with

❖ Better insight into the technology that supports the business

❖ Visibility to senior leadership

In addition to these, you may decide that you do enjoy the challenges you face on IT projects and decide to look for further opportunities to help drive change in organizations. With all of the challenges and opportunities facing organizations today (cybersecurity, artificial intelligence and big data to name but a few), the demand has never been higher for professionals who can help organizations structure and navigate the changes that come along with these challenges.

Takeaway: Realize that you can get much more than just "experience" from working on an IT Project. The relationships you build and the larger body of knowledge that you acquire can help you in future roles, whether part of another IT project or as you move through an organization.

Final Thoughts

Finally, I would like to hear from you!

It is my sincerest hope that you found this book informative, and that it provided you with a good foundation upon which to build your career success.

If you found this book useful (or found any mistakes in it) please drop me a line at steve@conceptia.com. Also, please consider writing a review on the site where you purchased it.

Be well and help the person next to you.

Steve.

Detailed Table of Contents

Acknowledgements

Thanks to the following people for providing support and input in developing this book:

Ms. Jesse Thomas, my second-grade teacher at Southmayd Elementary in Houston, Texas, who introduced me to my first computer, the Apple][e. She sparked something inside me that is still burning to this day.

My mother who bought me my first computer (also an Apple][e that I still have), despite the fact that we could not afford one at the time.

My father, who taught me how to work with my hands, to respect those that do and the value of a day's work.

My sister, Gerri Huck, for...well everything.

My wife, Monique, for being a constant voice of encouragement, for reminding me that there's more to life than work (and for keeping the little madam out of my hair when I needed to write!)

My daughter, Nyla, already a coach at the ripe old age of two, who taught me not only the importance of seizing the day, but how. She has also taught me how to keep the twin demons of self-doubt and procrastination at bay ("No!").

Andre Jaundoo and Chris Warren, friends who were there for me in a way that gave me the space I needed to complete this book.

Yamini Sharma and Amanda Kieval for your thoughts and feedback on an early draft of this book. It is surely a much more thorough book thanks *to your input.*

Sheila Heen, Douglas Stone and my fellow Business Book Writing Retreat (BBWR) attendees for the insight, guidance and encouragement offered during our time together. This book may not have seen the light of day had it not been for that week.

The staff of Castle Hill where BBWR was held and a large portion of this book was written. Your attention to detail and quality of service allowed me to forget my personal needs for a week and focus on this effort.

A big THANK YOU to my editor, and chief supporter, Lona Neves. One thing I've learned through writing this is that in addition to having someone responsible for cleaning up your words and making suggestions to improve readability, every author also needs someone in their corner propping them up, encouraging them to continue, and generally ensuring that the project continues to move forward. Lona was that person for me.

And finally, a special thank you to the young (and not so young) men and women I've tutored/coached/mentored over the years. I am truly honored to have been a small part of your journey and continue to be inspired by everything you are doing.

About the Author

As the founder and principal of Conceptia, Inc., headquartered in Houston, Texas, Steve Pinckney has helped dozens of organizations execute technology projects more effectively. He has over 20 years of global experience in project management, organizational change management, and solution delivery. His work has spanned the investment banking and oil and gas industries, as well as the government sector.

Steve is also an Executive Coach, specializing in developing high performing teams. Through a combination of individual and group coaching, he helps accelerate ROI by enabling project teams and their leaders to get clear on their mission and goals, identify and remove obstacles to their productivity, and improve team communication and interactions.

99246058R00068

Made in the USA
Lexington, KY
15 September 2018